CROWN LYNN

A NEW ZEALAND ICON

VALERIE RINGER MONK

PHOTOGRAPHY. STUDIO LA GONDA

PENGUIN BOOKS

CONTENTS

INTRODUCTION

'As long as there is a Crown Lynn dinner set on a New Zealand table or an NZR cup souvenired from a Taumarunui tearoom, Crown Lynn will endure, tangible reminders of our industrial history.' *WESTERN LEADER*, 28 JULY 1989.

MOST CROWN LYNN COLLECTORS seek out the trickle-glazed vases and animal figurines, or the hand-made pots by Ernie Shufflebottom and Frank Carpay and Daniel Steenstra, but in reality Crown Lynn's mass-produced household dinnerware is much more widely known. In its heyday in the 1970s the factory employed over 500 people and made millions of items each year. The majority of New Zealanders used Crown Lynn products every day. Admittedly, at the time there wasn't much choice; Tom Clark and his team lobbied successive governments to keep competition from imported ware to a minimum. At its best, though, the Crown Lynn product compares very favourably with English ware. Crown Lynn was one of the few companies to take design seriously in those early days, and as a result much of the mass-produced domestic ware is well designed and attractively decorated while also being durable. Even today, 20 years after the factory closed, older people hunt through piles of plates in second-hand shops, searching for replacements for their Crown Lynn dinner sets. 'Good value for money,' they say.

As I researched this book, it seemed that every second person I met had a Crown Lynn story to tell: Adrienne, who toured the factory

[Previous page] McAlpine jugs.
[Opposite page] Wharetana Ware wall vase.

as a girl and saw a woman with peculiar green nail polish, hand painting plates; Wendy, whose father and his Labour Party mates sat round the table discussing politics and drinking tea out of huge Crown Lynn cups; Kitty, who won her Willow pattern tea set at the Glenfield housie nights; and Mary, whose Ponui dinner set was considered exotic when she received it as a wedding present in 1972. Then there are the marae in Turangi and the public hall in the Hokianga, both of which have been using monogrammed Crown Lynn ware since the 1960s.

The expression 'New Zealand icon' is over-used, but for Crown Lynn it is accurate. Crown Lynn china was made in New Zealand, and the shapes were designed in New Zealand, as were many of the decorations. Through the 1960s and 1970s Crown Lynn dinner sets were routinely given as wedding presents, Crown Lynn whiteware decorated mantelpieces, and we all drank out of Crown Lynn cups and mugs. At one end of the scale were thousands of heavy, indestructible brown plates for hotels and tearooms, at the other end Frank Carpay's finely balanced and boldly decorated vases and platters that are worth a small fortune today.

Crown Lynn products were not always in such keen demand. In the early days the factory – then known as Ambrico – was something of a laughing stock as it produced thick yellowish speckled cups with handles that fell off. But in wartime they were all that was available, which gave the fledgling factory the break it needed to start producing a better quality product. The most impressive thing about the Crown Lynn story is that Tom Clark and his team never gave up. There were economic ups and downs, and more than a few technical disasters. There were times when they almost despaired, but somehow they managed to pull themselves together and get on with the business of making crockery. Tom Clark was one of the last of the 'makers', building up a large manufacturing business before the New Zealand economy was taken apart by David Lange's 1984 Labour Government.

It is the human connection that makes the Crown Lynn story so interesting to write. Over the 50-odd years of its existence, thousands of New Zealanders worked at Crown Lynn, and during my research I met many of them. Strangely, many of the Crown Lynn staff I interviewed for this book had very little of the product in their cupboards. Perhaps, like Gina Bird, daughter of decorator Doris Bird, they felt there was a never-ending supply. In the Bird household there was always the same calm response to a broken cup or plate. It was just 'mind your fingers' as you picked up the pieces. There was plenty more where that one came from.

Above all, the Crown Lynn factory is remembered as a family business. Most of the employees, especially those who started in the 1960s or before, have extremely fond memories of the place. People worked there for decades. Take Tony Rakich, who wheeled round heavy racks of greenware. The racks had to be pulled, not pushed, so Tony spent 40 or so years walking backwards. Tony is representative of the people who gave their all to Crown Lynn, working long hours, at often tedious and repetitive jobs, on barely acceptable pay rates. This book is a testimonial to their hard work and commitment.

SOURCES / When I started researching this book I discovered that there was no ready-made 'Crown Lynn archive' for me to delve into. Much of the early written information about Crown Lynn had been lost in factory fires over the decades, and valuable material was thrown out after staff changes and during those sad days when Crown Lynn finally closed down in May 1989. Thus, interviews with people became invaluable; but as we all know the human memory is fallible so I have backed up verbal information with written records whenever possible. Often those records are held by someone who used to work at Crown Lynn and in many cases they are the only known copy that still exists. A case in point is a staff newsletter from 1948 which was kept purely because an engagement was announced in it. This newsletter contains 32 pages of information which otherwise would have been unavailable to me – or to any future researchers.

Unfortunately, even printed material is not infallible, and it has proven impossible to accurately pin down some dates, especially for the early years. For example, different newspaper articles give different dates for the manufacture of the first railway cups and the bowls and mugs used by the American army. In those cases I have made an educated guess as to which date is most likely to be correct.

As I was researching this book I became aware of the huge array of people who were involved in Crown Lynn – colourful, hard working, innovative – each and every person who worked there contributed something. It is not the purpose of this book to record the names of everyone who was employed at the factory, and I know I have missed out some who played vital roles there. Some who worked at Crown Lynn may feel that too much emphasis has been given to certain individuals, and perhaps not enough to others. This, however, is unavoidable in a project of this type. Thousands of people passed through Crown Lynn in its 50-year history, and I cannot mention them all. Nor can I interview them all; interview subjects were chosen because they were representative of a group of other people who did similar jobs.

My research papers have been archived in the Waitakere City Council library. In addition, oral historians have recorded interviews with people who worked at Crown Lynn and those, too, are available from the library. This book is not meant to be a scholarly tome so I have not used footnotes. However, a manuscript noting sources of information has also been lodged at the library.

SIR TOM CLARK
1916–2005

'Tom Clark has spent his life heading for the sun and, whether there has been success or failure, he has lived every "goddam" mile of the way.'

JILL McCRACKEN, *NEW ZEALAND LISTENER*, 20 SEPTEMBER 1971.

WHILE I WAS WRITING THIS BOOK Sir Tom Clark died. Tom, of course, was 'Mr Crown Lynn'. He set up the original factory in the late 1930s and was closely involved in it for nearly 40 years. At our first meeting he told me that the Crown Lynn story wasn't about Tom Clark, it was about his team, and he sent me away with a huge list of people to interview. And interview them I did – but fortunately I also spent many hours with Tom as he told me his story.

It is easy to say that he was born with a silver spoon in his mouth, that his family owned a pottery and he stepped into a ready-made job. The truth, though, is not that simple. He was a difficult young man, a reluctant scholar who was 'redeemed' by being set to work at 14 at his family's brick and pipe works. He soon began making other things besides bricks and pipes, and the rest, as they say, is history. Looking back in his old age, Tom was astonished at his youthful drive and self-confidence. At no stage did he seriously entertain the idea that the factory would not survive, even in the late 1940s when markets collapsed and creditors were knocking at the door.

One of Tom's talents was the ability to gather an excellent team around him and to inspire, cajole and bully them into working harder and longer than they had ever expected. His catch-cry 'What's new today?' kept the entire workforce striving for improvement, especially during the early days when the team was relatively small and tightly knit. The people who worked for him – no matter at what level – say that his secret was his interest in everyone. He would stop and greet new staff, learn their names, and remember them the next time they met. He told me that Crown Lynn was nothing without the people who worked there: 'The team was quite outstanding. The people were unique. They were just bloody wonderful. Still, the thing that stands out to me is the personalities and the relationships that I had with those guys – not the various bits of Crown Lynn that we made.'

Tom was a complex character. He could be loud and foulmouthed – to the extent that hand decorator Eileen Machin's boss protectively sent her out 'for a walk' when Tom visited their workshop. But he was a perfect gentleman when he was lobbying members of Parliament or showing the Queen around the factory. He could hobnob with the rich and famous, but he also partied with the factory workers. Above all, Tom was proud of his achievements and proud of the achievements of those who worked with him. He was delighted when I brought him a hot-water bottle he'd made in the 1940s and similarly delighted when we talked about one of the first cups ever made in his factory. It was thick and heavy but the glaze was still clean after 55 years – 'It would make a good missile, wouldn't it?' he said.

He was direct in his approach to others, and could be blunt to the point of rudeness. He was not afraid of sticking his neck out. And, most important, he was not afraid of hard work – 'head down and backside up' was one of his favourite expressions. In his younger days Tom always had a raft of projects on the go at any one time. Once Crown Lynn was up and running, his mind zigzagged off into other projects – prefabricated building panels, ceramic

granules for roof tiles, fire-retardant doors, abrasive granules, synthetic fibreglass vanity basins. With all these projects Tom saw an opportunity and jumped in. And, on the side, he was a racing driver, owner and skipper of ocean-going racing yachts, and, later, supporter, friend and mentor of America's Cup skipper Peter Blake.

Especially during those early days, Tom was obsessed with Crown Lynn. Even on his honeymoon with his wife Trish, he would stop the car, leap out, pick up a piece of clay from the roadside and put it in his mouth to test for grit. Good clay, he said, is like butter, with a smooth, grit-free texture. For 20 years he was overseas for months at a time almost every year. When he was home he was often at the factory from six in the morning until seven or eight at night. And he expected the same commitment from his managers. There is no doubt that company wives and children often had a tough deal. He could be generous with valued employees, but in return he expected 100 per cent commitment to the job in hand, and was disappointed when anyone failed to live up to his expectations. 'Get on with it, sonny' was another favourite expression. Tom had a large family and, in later life, expressed regret that his intense involvement in his work had taken him away from New Zealand for such long periods.

Almost larger than life, Tom was tall, energetic and highly intelligent. His staff remember him as a person who could converse on any topic; usually he knew something about it and if he didn't he would listen carefully until he did. He had a short attention span; he would think up a new project and get it going, then hand it over

to one of his assistants while he moved on to the next great scheme – usually successful, sometimes not. He hated making mistakes and instilled in his team his own passionate drive to get things right. The famous problem of the non-sticking cup handles almost drove him mad, and his engineers and ceramicists toiled long and hard until the problem was solved.

Tom Clark had more drive than most of us will ever have. When I first met him at the age of 88, he was often out in the garden, mowing the lawn or pruning his roses. When he became too unwell to work outside he started learning how to use a computer and was making excellent progress on it before he died. He was a man of enormous courage and energy. I hope that his Crown Lynn legacy lives on for generations to come.

VALERIE RINGER MONK May 2006   **09**

40s

THE EARLY DAYS
1935–1948

'It could be said that Crown Lynn crockery ruled supreme for 50 years on the dinner tables of the nation.' *Western Leader*, 28 August 1989.

In West Auckland, tucked behind the Lynnmall shopping complex, there is a cluster of little residential streets. One is called Crown Lynn Place, another Clark Street, another Ambrico Place. Apart from these names, there is little to suggest that on this site once stood Crown Lynn, a bustling, dusty, hot, noisy, sprawling factory that employed over 500 people and turned out 15 million pieces of household china in a single year.

THE BEGINNING / The origins of Crown Lynn can be traced back to the 1890s when Rice Owen Clark bought a farm at Hobsonville, west of Auckland. To drain his soggy land he created drainpipes from the local clay. Family legend has it that he wrapped clay around a section of tree branch, burned the whole lot on a fire, and presto, there was a drainage pipe – rough but effective. As his manufacturing methods improved, neighbouring landowners began to buy his products and in the early 1860s he had a thriving pipeworks.

By 1906 the factory made salt-glazed garden pots and urns, bread pans and storage jars as well as bricks and pipes. In the 1920s the Clark family business joined forces with several other potteries in Auckland and Wellington to form the Amalgamated Brick and Pipe Company. The Clark plant at Hobsonville shut down in 1925 and production was centralised in New Lynn, which offered better clay, plentiful labour and a rail siding close by. In 1928 the

The Amalgamated Brick and Pipe factory in 1929.

factory was producing one million bricks a month.

By 1931, though, the business was feeling the effects of the Depression. The workforce of around 250 shrank to only eight as unsold bricks and pipes piled up. The company's managing director, Thomas Edwin Clark Snr, was also knocked by the Depression. He could no longer afford school fees, so his son Tom, a boy of 14, was plucked from school and set to work. T. E. Clark Snr believed that hard work would make a man of the rebellious lad and soon had him labouring around the factory and digging in the clay pits. The digging was all done by

hand, by hard men with muddy trousers hitched halfway up their legs and tied below the knee with pieces of string known as bowyangs. It was tough, messy and sometimes dangerous work and all his life Tom Clark remembered the hard toil of those days – 'head down and backside up' – with heavy, sticky clay clinging to boots and clothing. His workmates taught him techniques that made the job easier. Clay slid off a wet shovel easier than a dry one, so water was always at hand. A grindstone was used to cut down new shovels to an effective size – 'A good guy with a little shovel will outshovel a big guy with a big shovel any day.'

After several years' hard slog in the clay pits and around the factory, young Tom had learned just about all there was to know about making bricks and pipes. He knew how to dig and mix clay, and how to mould bricks and pipes. He could load a kiln, fire it and unload the result. He knew all about salt glazing sewer pipes: salt was thrown into a hot kiln where it volatilised and coated the pipes with a brownish glaze as they cooled. The salt came from the local tannery, a waste product from tanning animal hides to make leather. By no means pure, it sometimes included pig's ears and other unsavoury items; but it was cheap and it served the purpose.

TILES AND ELECTRICAL GOODS / After learning all he could at the pipeworks, Tom Clark spent a few months at the company's brickworks in Kamo, just north of Whangarei. Back at New Lynn, his mind began ticking over. He felt that the factory could make more advanced products than bricks and sewer pipes, which were highly vulnerable to the economic swings already seen in the Depression. 'I was convinced that we could do more imaginative things with New Zealand clay than just pouring it into a great big mixer and squirting it out of a die.' Barely 20 years old, he began studying books and experimenting with different clay mixes and treatments to produce smaller and more specialised items.

He saw an immediate market for acid-resistant tiles, which were needed for the walls and floors of dairy factories and abattoirs. Like most goods requiring technical skills and specialised machinery to manufacture, they were imported from England. Some factories made do with concrete – a vastly inferior option.

Tom Clark needed funding to implement his vision and took a formal proposal to the Board of Directors. He was granted £5000 and with the aid of Ray Ockleston, the son of another brick-making family, he began attempting to make tiles. After much experimentation the two young men discovered that the best tiles were made by dry-pressing. A precise metal die was filled with a finely ground dry clay powder, and massive pressure applied to consolidate it. The resulting hexagonal tiles were fired in a kiln that the two men had designed and built. Clark and Ockleston were both young, driven and excited by their new projects, but neither of them had any experience in the production of fine clay products. They asked questions, studied textbooks on the subject and learned by trial and error. Many years later a company publication described this period – between 1935 and 1938 – as 'two men feeling their way diffidently in a dark corner'.

Before the pair could go into full production, they had to invent or adapt several machines. They built a pug mill to grind up the clay and blend it into a smooth paste. A second-hand filter press from the local Chelsea sugar works was adapted to squeeze out excess water, and then the cakes of clay were air dried before being milled into a fine powder ready to be pressed into shape and fired. This was hot, hard work. The machinery for pressing tiles was hand operated, and only made one tile at a time. The coal-fired kiln needed to be tended and stoked regularly for up to 48 hours. Inevitably there were disasters. Once they fired the kiln too quickly and its precious contents came out misshapen and useless. Another time the tiles all melted together. After that they were separated with silica sand during firing. These mistakes were intensely frustrating, particularly as so much manual work went into making each individual tile.

Despite the setbacks the men successfully made and sold their product, and after about 18 months other opportunities opened up. In the early 1940s thousands of New Zealand homes were being connected to electricity for the first time. There was a growing demand for radios, stoves and hot-water cylinders, each of which had porcelain components. Soon the little factory was making radiator bars, ceramic spools for hot-water cylinder elements, ceramic stove elements, parts for radios, and insulators for electric fences and power poles. There was also a steady market for non-skid tiles for the outside edge of steps. Progressively, through that first decade the team expanded, along with the product range. By 1940 there were six men, among them Vern Gray, a skilled engineer hired to make the machinery needed to mould the small and precisely shaped electrical components.

THE FIRST HOUSEHOLD WARE / Near the end of the 1930s Clark and his team took their first tentative steps towards making domestic ware. Until now they had used only one manufacturing process – dry-pressing.

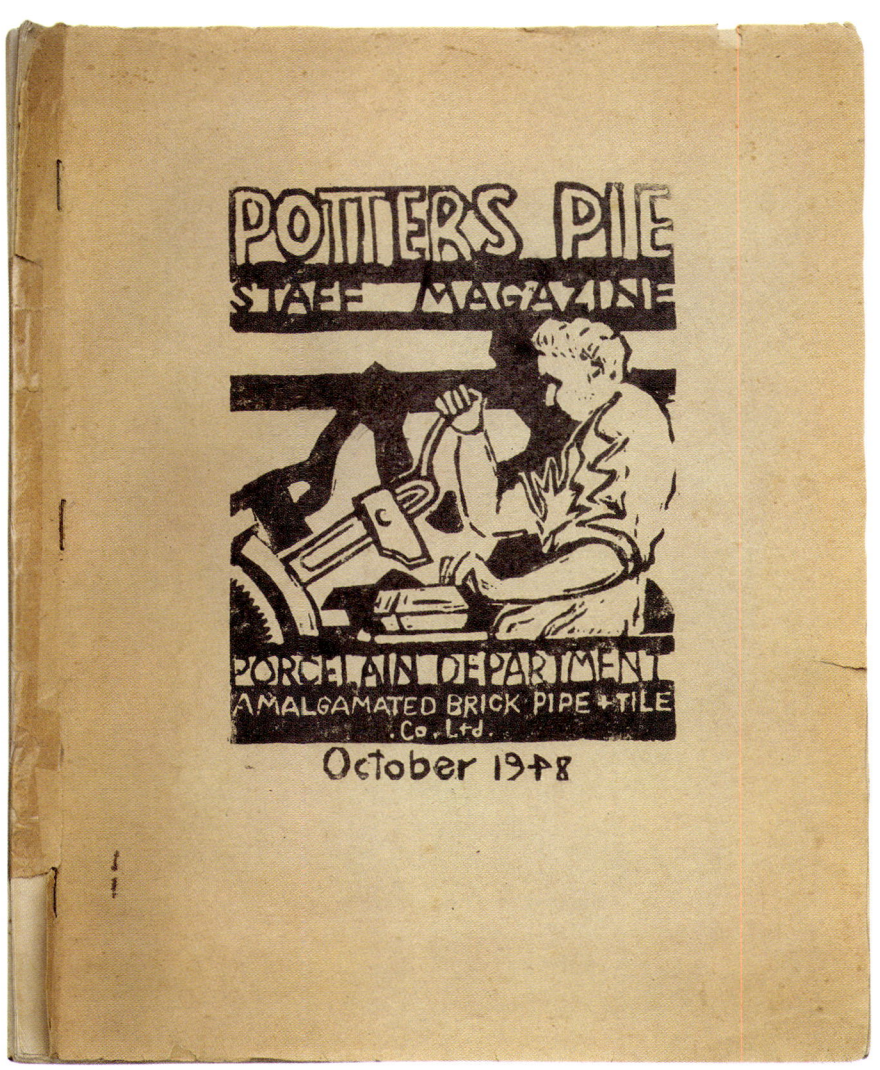

The cover of the October 1948 staff magazine shows a man using a hand-operated jigger. The illustration is probably by David Jenkin.

Making household items required completely different skills and they began to experiment with slipcasting, using plaster of Paris moulds to form shapes from liquid clay. In another vital breakthrough they managed to make their first very basic glazes.

By now the little department had outgrown its first premises, a room at one end of the pipe factory, and a two-storey building, later known as the 'Old Factory', was erected alongside the pipe factory. The clay-preparation machinery was on the ground floor, and the laboratory, workbenches and office were upstairs. Fortunately, the expense of building the factory, buying tools, making machinery and hiring staff was balanced by steady profits. New Zealand was emerging from the Depression and there was money about. The fledgling electrical industry was booming and the new enterprise, by then christened the Porcelain Specialties Department or Specials Department for short, was already making money.

THE WAR / Production was still small-scale and experimental, but this changed dramatically when the Second World War began in 1939. At the time New Zealand's crockery was imported from England, but by 1940 all available shipping capacity was being used to transport troops and army supplies. New Zealand faced a severe shortage of household plates, cups, jugs and bowls. The only possible domestic manufacturer was the small factory in New Lynn. To meet wartime demand Clark and his engineers had to speed up production. Slipcasting was too slow, so they developed a primitive hand-operated jigger to shape cups and plates. The operator pulled down on a lever to press a piece of soft clay between a template and a rotating mould.

Many years later long-time staff member Bill Wiseley recalled that there was huge excitement in the factory when they managed to produce the first jiggered cups, even though they were heavy and ungainly.

By 1941 jiggering was in full swing and a big new kiln had been built. This allowed for a massive expansion in production, which by this time included kitchen bowls and porridge plates as well as cups.

'KING OF THE KILNS' / In the first few years the factory used an assortment of kilns, either coal-fired or diesel-fired. Some were inherited from the brick and pipe factory and others were designed and built by Clark and his engineers. Like most kilns they were temperamental, and it required much trial and error and considerable skill to get the best out of them. This job fell to Fred Hoffman, the 'King of the Kilns', who later held the position of factory manager. A loyal lieutenant throughout his 30 years in the business, Hoffman was almost always at work by 6.30 in the morning. However, his fondness for other pursuits sometimes got him into trouble. Clark recalled that in the early days Hoffman would sometimes nip across the road for a quick game of billiards at the local saloon during his lunchtime. He was skilled at the game and soon a pile of money – much more than his weekly wages – would build up on the table. He would lose track of the time until he was dragged back to work. On the job, though, he ruled with an iron fist. Tom Hodgson, who worked for him

One of the first cups made on a jigger. It would have been manufactured during the war, probably for the New Zealand market.

15

in the 1950s, recalled that latecomers were told they were stealing from the firm: 'If you were two minutes late he got into you like a ton of bricks.' Hoffman was not above risking his own safety for the good of the company. Through the 1950s and 1960s, if there was a problem inside a kiln, the slightly built Hoffman would be bundled up in heat-resistant clothing and sent into the red-hot tunnel on a wheeled trolley. The aim was to clear the problem without losing valuable time by stopping production entirely.

THE AMERICAN ARMY / 'Crown Lynn today is quick to point out that the ware was made to specifications laid down by the Americans and was not their design. Thousands were produced for the Pacific fleet, but not for submarines because, as one wit at Crown Lynn suggested, "they would have sunk under the weight".' New Zealand Ceramics, July 1966.

By 1943 the Second World War was well under way and the American army had thousands of soldiers based in New Zealand and the Pacific, all needing plates and mugs. A friend at the New Zealand Government Stores Board alerted Tom Clark to the opportunity, and he immediately contacted the American army and set up a meeting. And that's how the immaculate 'Lieutenant-Colonel somebody-or-other' came to visit

Clark in his rough workroom. 'I remember to this day how embarrassed I was bringing him round this area with clay dust on the floor everywhere. I had to sweep off a bit of space on a chair for him to sit on.' Despite the primitive conditions Clark managed to persuade the army officer that he could do the job, and the next day he returned with specifications. The bowls and mugs had to be thick and solid and made of vitrified porcelain, which is fired at a higher temperature than earthenware and is much more robust. The Lieutenant-Colonel wanted thousands of these items and Clark took a huge gamble when he said he could make them on time.

This order was the trigger that got the pottery up and moving. Clark recalled that his team had been thinking that 'one day, one day' they would mass-produce vitrified ware, and once the Americans entered the picture they had to do exactly that – and do it smartly. The problem was that there were huge gaps in their technical knowledge and the factory was not set up for mass production. In a very short time they needed to develop a vitrified clay body, set up enough hand-operated jiggers to make the bowls and handle-less mugs and build enough kiln space to fire them in. Then, said Clark, 'We produced thousands of those things. Repeat, repeat, repeat. By hand.' They were thick and

heavy, made with a putty-coloured body and finished in a greenish-brown or transparent glaze. A few, possibly experimental, were dark blue.

They were not pretty, but they were made on time and to the Americans' specifications. They had to be able to withstand harsh treatment; the story goes that on the ships they were literally shovelled into giant dishwashers. After the war an American army expert pointed out that it wasn't such a bad thing that the cups lacked handles. As the ships pitched and rolled the handles would have been the first thing to break.

Tens of thousands of these mugs and bowls were produced, to be used by the American army in New Zealand and the Pacific. They are now collectors' items – if you can find one. Thousands would have been broken by rough handling and thousands more were probably dumped after the war.

THE FAMOUS RAILWAY CUPS / The American army was not alone in its need for crockery. The New Zealand Government had lost access to English-manufactured ware and crockery was badly needed for the railways, armed forces and hospitals. Around the time that the little factory began producing ware for the American army, it also started making cups and saucers for the New Zealand Government. The most well-known product

17

A new role for railway cups. In the early to mid-1960s a slightly larger version of the railway cup shape was used for a completely different purpose. Upturned and without a handle, the shape made a perfect 'umbrella' for electrical fittings on power poles. Crown Lynn's sister company Technical Ceramics created a fuse which fitted neatly inside the cup, keeping electrical connections clean and dry. Throughout New Zealand these cup shapes can still be seen on older wooden poles.

of this time is the legendary railway cup.

The first railway cups made at the factory were very primitive earthenware, yellowish in colour and coated with a clear glaze. A huge 'NZR' was stamped on one side. At this early stage Clark hadn't managed to attach handles. As he later explained, 'Why were they made without handles? For the best reasons in the world. One, we didn't know how to make handles; and two, even if we had known, we didn't know how to stick the damned things on.' Within a year or two the factory was making jiggered cups of vitrified porcelain, complete with handles. The first 'block handles' had a much greater surface area attached to the mug than the later 'ear-shaped' versions which were attached only at either end. Whatever the early cups lacked in elegance, they made up for in solidity. 'If one hit you on the head it would probably kill you,' said a staff member years later.

Several different variations of the railway cup were made over the years, all bearing the New Zealand Railways logo. In the late 1960s and early 1970s Crown Lynn made cups and dinnerware with the railways logo zigzagging around the top. Today the classic Crown Lynn railway cup is one of New Zealand's enduring symbols, appearing on postcards, in Kiwiana books and even on a postage stamp.

Although Crown Lynn is traditionally identified with the railway cup, Temuka Potteries in the South Island also produced cups and saucers for the railways in the 1940s. And not all railway cups were made in New Zealand. Before the Second World War the railways used cups imported from England, and much later, probably from the late 1970s onward, English-made ware again made an appearance.

WARTIME GROWTH / During the war Australia, too, was completely cut off from English imported china. Arthur Martin was hired to find new markets, and before long almost half the factory's production was being sold across the Tasman. This was a period of great innovation. If the small team thought there was a market for an item, they would work out how to make it. The factory was more than a full-time job; it was an obsession. Years later Clark joked that he spent so much time at work that he had a bed next to the kiln. By the mid-1940s the factory was manufacturing jiggered cups, plates and bowls, and slipcast hot-water bottles, chamber pots, shaving mugs, vases and electric jugs. There was also a strong market for ceramic moulds for making rubber goods such as condoms, rubber gloves and teats for babies bottles. Kitchen bowls were also a popular item – in this era everyone made their own biscuits and cakes. Creating each new product was a challenge, and the team were delighted when they got it right. Over 60 years later Clark was genuinely proud of those early items. He described the hot-water bottle as a benchmark, a collectors' item. 'We thought we could make something nice that had a bit of design about it. We could cast it, we could fire it. I tell you, that is really something.' The hot-water bottles were made of a special strong, waterproof vitrified clay body, which was unstable during firing. There was much trial and error before they managed to consistently produce bottles that

Railway cups from the 1940s and 1950s. Many New Zealanders still remember the hasty cup of tea at Frankton Junction and again at Taumarunui on the Limited Express train trip from Auckland to Wellington. Both stations had cafeterias where you could buy a slab of yellowish fruit cake or a thick ham sandwich and a cup of over-brewed milky tea. The cups were white and chunky, with the New Zealand Railways logo emblazoned in blue or black. The queues were long, the tea-pouring slow, and often the last of the travellers didn't get their cuppa until just before the 'all aboard' whistle blew. Then there was nothing for it but to sprint to the train with cup and saucer in hand, and settle down on the well-worn leather seats to drink your tea in the lurching carriage. When you were finished, you either put the cup on the floor, put it in your bag to furnish your student flat, or you tossed it out the window. Over the years throwing railway cups out the train window became something of a tradition. Travellers saw it as a challenge to try to break the cup by tossing it as hard as possible at a rock or a tunnel wall. Now, decades later, treasure hunters scour the railway lines in search of jettisoned railway ware hidden in the grass and scrub. To university students, railway cups and plates were fair game as free crockery – and back at the student flat, after a few beers, cup-smashing competitions were great fun.

didn't sag out of shape in the heat of the kiln. And there was more trial and error before the screw tops – also ceramic and fitted with a rubber washer – were made to fit well enough not to leak.

AMBRICO WARE / By the mid-1940s Porcelain Specialties Department no longer seemed an appropriate name for the busy and productive factory, and the company received a real name. Without giving it much thought Clark called his new products Ambrico Ware, a name made up of the initials of the parent company Amalgamated Brick and Pipe Co.

CLAYS AND GLAZES / In the fledgling factory different types of clay 'body' had to be developed before the new products could be made successfully. There was a need for an earthenware body for domestic ware, and a harder, more durable vitrified porcelain for hospitals and hotels. The body could be a combination of at least nine different clays and other materials, all mixed in exactly the right proportions and to the right consistency. It took a great deal of research and trial and error before successful mixes were developed. These experiments had little scientific basis until 1943 when the first chemists, Denis McClure and George Peters, were hired. Four years later there were 12 laboratory staff. The laboratory was under huge pressure to formulate a clay body which was durable, reproducible and consistently turned white when it was fired. Thousands of samples were tested; thousands of combinations were mixed, fired, then rejected. Some of the early ware was yellow and some had greyish speckles caused by iron pyrites contamination. Much of the clay used in the early days came from Whau Valley near Whangarei. This was the beginning of a decades-long battle to make the factory's product thinner and whiter. At least until the end of the 1940s the factory was still making some yellow ware, partly because some householders preferred it to the rather muddy coloured white items.

At the same time as they were struggling to create an acceptable clay body, the laboratory staff were developing glazes. Long-time glazing expert Ron Absolum recalled that when he joined the factory in 1946 crockery was plain white – 'and not too white at that' – and finished with a clear glaze. The only colour used was a touch of cobalt blue on decorative vases. The chemists soon learned how to add metal oxides – lead, cobalt or copper – to create

A hot-water bottle made by the Specials Department in the mid-1940s. A cylindrical version was also produced.

20

different colours. Tom Clark's aunt, Briar Gardner, helped her nephew considerably during those early days. One of Auckland's first art potters, Gardner passed on her knowledge of glazes and in return 'borrowed' clay when she ran short, and even enlisted factory workers to help her with heavy work.

A HOUSEHOLD ESSENTIAL / In the days when most houses lacked inside toilets, the humble chamber pot was a big seller. The pot was kept under the bed and emptied in the outside 'long-drop' toilet every morning. Over three decades Crown Lynn made at least seven different shapes of chamber pot, including a small child's model and miniatures in various colours. Some were embellished with flower motifs. As late as 1969 chamber pots were still being made. For many years hundreds of pots, many made by Crown Lynn, festooned the ceiling of the now-defunct Wagener Museum in Houhora in the Far North.

ELECTRIC JUGS / Between 1944 and 1949 the factory made electric jugs for heating water. At first the heating elements were not insulated; thus, if the jug was turned on, the water was 'live'. The jug itself had excellent insulating properties, which made it safe to touch. Later, stainless steel safety elements were installed. The electric jugs were made in a variety of shapes and sizes, and sold in New Zealand and Australia. Thousands were sold through the Farmers Trading Company. Despite their limitations many householders found these early jugs preferable to turning on a sluggish solid stove element or lighting the coal range. Before long, ceramic electric

jugs went out of production as cream and green enamelled metal jugs took over the market.

THE PARIS DINNER SETS / The first dinnerware made by Ambrico was the distinctive ridged Paris design. The first Paris plates were made around 1943, using a yellowish earthenware body shaped on hand jiggers. The jigger moulds were ridged and hence the plates and cups were too. Later, the same pattern was manufactured on a stamping machine, then on the English

The decorated chamber pot was a standing joke at Crown Lynn. When warehouseman Harry Bird left to sail to the Pacific Islands with his son Cliff, he was offered farewell drinks from a pot inscribed 'Wash me well and keep me clean & I'll never tell what places I've been'. In this photo his granddaughter Donna looks on as Harry drinks from it. Another recipient was presented with a chamber pot decorated with floral transfers – and a large black staring eye, dead centre on the bottom.

machines which were imported after the war. Jugs were slipcast. The first cups had block handles, and later ear-shaped handles in various styles were fitted.

TRIAL AND ERROR / Well before the 1940s the English, German and Scandinavian potteries were making exquisite fine household ware in bone china, earthenware and vitrified china. In comparison Ambrico's products were heavy, ugly and badly made. But the miracle is that they were made at all. In those early days everything was a challenge. Even designing and making a jug had its pitfalls. How to shape the jug so it could be easily cleaned inside? How to make it durable and appealing to the eye? How to shape the spout so it wouldn't dribble? How to make a clay body that didn't crack, craze or sag out of shape in the kiln? Clark had a good knowledge of ceramics after his experiences with bricks and pipes in the family factory, but it was a huge leap from those relatively primitive processes to producing household ware. Throughout New Zealand other potteries were struggling with the same problems, and as workers moved from one factory to another there was some cross-fertilisation of ideas. Apart from that, all Clark's information came from books – which he treasured until they were destroyed in a factory fire in 1956. At first Ambrico had

no quality control to speak of and products were sold after a cursory check. Then, when retailers began to send back cracked wall vases and badly crazed cups, a testing laboratory was set up to ensure that the finished product was up to standard.

FINDING MACHINERY / As the factory expanded machinery was a big issue – how to find workable equipment at an affordable price while wartime restrictions limited imports. When a machine was needed to grind up raw clay, the engineers searched far and wide for something suitable. Finally, in the Coromandel goldfields, they found two abandoned ball mills which had once been used for breaking up gold ore. The rusty hulks were hauled back to the factory at New Lynn and rebuilt. Once returned to working order, the 5-metre-long iron cylinders were

(Above) This small ashtray, probably from the 1940s, shows obvious manufacturing flaws. The glaze has bubbled and the clay body has split during drying or firing.

(Opposite page) Paris dinnerware.

Tom Clark in 1944 with two of his factory staff. Many of the workers of that time were older women whose husbands were in the army overseas. The photo was taken at an Ambrico social event.

lined with ceramic tiles and loaded with river stones, then filled with wet raw clay and set slowly revolving. In time the stones ground up the clay into a smooth paste. These ball mills were so successful that one was still in use when Crown Lynn closed in 1989. The factory also used tube mills, which were left revolving slowly all night to mix the clay to the right consistency. For many years the clay processing and mixing machines were loaded by hand. Different types of clay were kept in separate heaps on concrete slabs or in bins, and men shovelled measured amounts of each into the huge mills.

During the war it was impossible to get reliable supplies of fuel oil, so Vern Gray and Ross Brewin designed and built a plant which produced gas from coal. Problematically, one of the by-products was heavy tar, which was poured into a pit. Every now and again the tar would be set alight and, recalled Brewin, the resulting cloud of smoke would black out New Lynn. 'The council used to threaten to sue us and we would say that we would never do it again, then in a few months we would.'

WOMEN IN THE WORKPLACE / As the war gained momentum, most able-bodied men joined the army, and women were hired to keep the business operating. Clark had no idea what facilities were required, so he asked Briar Gardner to help and on her advice built new toilets and a lunchroom. Among the first women was a clerical assistant, Jean Hollinger, who soon became engaged to Ray Ockleston and left work when they married. Her successor was 22-year-old Melva Fremlin who started in 1942. Before Melva was allowed to begin work, her mother inspected the premises to make sure she would be safe. Melva was responsible for wages, administration and making tea. The phone system was extremely basic and Melva remembered racing through the busy, noisy factory to find her boss every time he was wanted on the phone. The place was very dusty and dirty, and the young women soon learned to wear clothes that didn't show the grime. After five years Melva left work to marry Ray Ockleston's brother Stan.

Women were also hired to work on the factory floor. One of the first was Bebe Woolright (later Cowdery) who began in 1942, at the age of 16. Her job was to pack tiny insulators in boxes and thread heating wires into ceramic stove elements. Later she began working in the office. To earn extra money some office staff worked in the factory after hours, making bowls on the jiggers. The clay was coated in oil to make it spin smoothly against the jigger, creating a disgusting smell as the oil heated up with the friction of the machinery. Jiggering was hard work in an incredibly noisy environment, and the women often had sweat dripping off their noses as they heaved around lumps of clay and hauled down on the jigger levers. For protection they wore heavy leather aprons. The women also got the job of packing the finished products – not a pleasant task as the hay used for wadding often contained blackberry thorns and animal dung.

In those early days women didn't find it easy to work in such a rough environment. It was a good 30 years before sexual harassment became a workplace issue, and one worker recalled that her first weeks at work were fraught with anxiety. 'I spent all my time blushing. We girls didn't want to go to the toilet because it meant going through all these men. You would try to hang on and hang on until the bell rang and the men had gone elsewhere.' Even if they managed to avoid a barrage of wolf whistles from the men's workroom, the young women weren't always secure in the toilet itself. The factory floor was made of timber spaced so wide that the men could stand underneath the toilet cubicle and poke sticks up through the gaps in the floor. However, the job had its good times too. The older women – among them stalwart Daisy Savage – took a protective interest in the young girls, and the young male apprentices used to walk the women home if they had been working after dark. Despite growing pains the female workforce continued to expand, and by late 1942 the factory employed 36 women and eight men.

By this time there were more than enough women to make up a netball team and to organise picnics and other social events.

THE EARLY VASES / By the early 1940s the factory was making vases and ornaments as well as utilitarian kitchenware. Mould makers created some elegant shapes – and some not so elegant – in quite large numbers. In the first two years around 70 different vase shapes were made, most by self-taught mould maker Jack Aberly. Later, Harry Hargreaves and Tom Stewart added to the range. As well as the more traditional shapes, variations on clogs and shoes were made, and a series of small vases were shaped in imitation of miniature logs. Many of these original vases were

similar in shape to hand-turned products produced in the 1930s and 1940s by Briar Gardner. These little vases sold surprisingly well. The country was starved of ornaments after years of wartime import restrictions and a rather oddly coloured little vase on the mantelpiece looked better than a jam jar full of flowers.

The vases are fairly thick, but many appear to be quite delicate because their rims had been thinned down after they came out of the moulds. The first vases were decorated in opaque glazes in pale blue, green and fawn. However, under the influence of designer David Jenkin, who was hired in 1945, Ambrico soon hit upon trickle glazing as a cheap and effective way of decorating the vases and the small

A sociable Ambrico staff picnic at Bethells Beach in 1946.

model animals which were produced at around the same time. The vase was coated in a base glaze, then another colour was trickled over the top. Sometimes a third colour would be added. Although some of the first efforts were not especially attractive, decorators soon became expert at this technique and produced appealing colour effects.

Many of these early vases have unusual marks on the base, either written in pencil or scratched into the surface of the clay. Marks include the date of manufacture and/or fractionated marks – combinations of numbers and letters, for example ES/31. These symbols were codes recording the type of glaze and clay body used and, sometimes, the way the item was fired. This meant that successful experiments could be repeated. Notes relating to the meaning of the individual fractionated symbols went up in flames in one of the factory's big fires. Nevertheless, the marks are quite distinctive and a reliable way of identifying Ambrico Ware and early Crown Lynn.

THE MODEL ANIMALS / In the 1940s Ambrico made its first small animal figurines. Many of the earlier animals are roughly made, clumsily shaped and finished with obviously experimental glazes. Early examples include a pig, a rabbit, a kiwi, two versions of a lion, and two elephant shapes, and egg cups in the shape of a bulldog, rabbit, frog and chicken. In those early days anyone who showed an interest was allowed to model small items and put them through the kiln. Thus there is quite a range of animals

(Opposite page) Small vases like these, ranging in height from eight to 10 centimetres, were made from the 1940s to the 1960s.

(Above) A selection of larger vases made by Ambrico. Most were traditional styles but the designers also experimented with free-form shapes. The two tall vases in this picture are 19 centimetres high and the shorter one is 10 centimetres.

SALISBURY WARE / In 1942 Ambrico found a promising new market for its growing range of decorative ware. The marketing expert Arthur Martin set up a partnership with Owen Salisbury and the pair began to decorate unglazed ware with paints. The earthenware pots were spray-painted then hand painted with flowers or other motifs. Sometimes decorative transfers were applied. This process was cheaper and certainly easier than glazing and allowed for a greater range of decorative styles than Ambrico was capable of at that time. For Clark this was a godsend. At the time, he was able to produce presentable shapes, but struggled with decoration. He was delighted when someone else took that headache away. Unfortunately, this profitable market proved short-lived. In 1946, when wartime restrictions ended, Salisbury Ware couldn't compete with similar imported ware and the factory closed down. However, by the mid-1950s Salisbury was back in production and Tom Clark (now trading as Crown Lynn) was again asked to provide unglazed blanks. During its heyday Salisbury Ware sold thousands of painted items all over New Zealand. Unfortunately, 60 or 70 years later the paint has begun to flake off and it is not easy to find a good example. It can be difficult to identify Salisbury Ware made by Ambrico. Salisbury bought biscuitware (the first firing of raw clay) from other potteries as well, so it cannot be assumed that all Salisbury Ware blanks were made by Ambrico. It is possible to make an educated guess as to the origin of a Salisbury Ware piece by comparing its shape with ware from Ambrico or Crown Lynn. For example,

The early animals (above) were quite primitive, but later examples (opposite page) were more detailed.

shaped with varying degrees of artistic ability. One staff member described the animals as terrible designs, decorated with glazes 'like a cat that's lost its way'. Later, throughout the 1950s, the animals were made in quite large numbers. The mould makers became more skilled, kilns and glazes improved and the animals were more elegantly shaped and attractively finished. Animals made in the 1950s and 1960s included various Bambi fawn figures, dogs in different poses, a fox, kingfisher, parrot, sea lion, lion and prey, monkey, seagull, duck, petrel, hippopotamus, horse and lamb. Many of these animals were unmarked, but others bear the fractionated marks which were used when new glazes or new processes were being tested.

a trickle-glazed version of the vase pictured opposite was also produced by Ambrico.

To compound the difficulties of identification, painted ware similar to Salisbury's was imported from England. Salisbury Ware generally bears no backstamps and can only be identified by its appearance or by small labels which were stuck onto the finished product – most of which have long since peeled off.

HARWYN POTTERY / During the war New Zealand was starved of ornaments, which had traditionally been imported. Arthur and Alice Partridge spotted an opportunity and bought glazed vases from Ambrico. Under the brand name Harwyn Pottery they hand-painted the glazed blanks, mainly with sprays of flowers. Since the paint was applied on top of the glaze the designs flake off easily and many examples show obvious signs of damage. It seems that after the war, sales dried up and the remaining stockpile of blanks was used to fill in an air-raid shelter.

THE FIRST AUTOMATION / By the mid-1940s Ambrico had again outgrown its premises and began building the 'New Factory', which was constructed mainly out of recycled wood from the old Clark brickworks at Hobsonville. With the increasing demand for Ambrico's products, the factory needed to produce more

goods in a shorter time. There was a limit to how fast a manually operated jigger could turn out cups and plates and saucers; a machine was the next step. Because the war was still on, the engineers couldn't visit English potteries to find out how things were done, so Vern Gray and Norm Stevens were dependent upon their own imaginations. And thus, in the mid- to late 1940s, was born a complex gadget called the 'automatic hot press multi headed stamping machine'. The machine could make four items at a time and was theoretically capable of making 15,000 pieces a day – 'Our dream of mass production,' said Tom Clark. The machine stamped out flatware with gas-heated brass shapes which pressed down onto a piece of clay placed in a mould. It was used to speed up production of the Paris dinnerware which was already being made on hand-operated jiggers. It could make smaller plates, cups, cereal bowls and saucers, but larger Paris items were still made on the jiggers.

Despite the engineers' best efforts this machine was always prone to disaster. It broke moulds with frustrating regularity, causing far too many stoppages and delays. After battling for some years to make it work properly, Tom Clark went to England after the war ended to take a look at the latest developments. He realised that the home-made creation would never be as good as

the English machines, so he brought back new semi-automatic machines for making cups and plates. Once the new equipment was installed the ungainly home-made machine was scrapped. In retrospect it was doomed from the start because it was too complex. The engineers had tried to build one machine that made everything from cups to saucers to cereal bowls. The English machines were much simpler to build and operate because they were specialised – one made cups, another made saucers and so on.

A NEAR DISASTER / In 1946 and 1947 a new 70-metre tunnel kiln, designed by Tom Clark and Vern Gray, was built by a team of bricklayers. Kept going round the clock, it fired mainly biscuitware. The ware was packed in saggars (protective ceramic cases) on specially made cars which moved slowly through the kiln on metal rails in a 48-hour cycle. This new kiln allowed a massive increase in production, but it had not been in action for very long when it very nearly came to grief. During the Christmas break Clark called into work to check that everything was going smoothly. Peering through an inspection hole, he saw to his horror that instead of moving steadily through the kiln the cars had jammed near one end, leaving the centre unoccupied. Under normal circumstances

firebrick shields fitted to the cars protected the concrete floor from the fierce heat at the centre of the kiln. Since the mishap had prevented the cars from reaching the kiln's centre, the unprotected part of the floor was glowing a dull red. The burners were shut down and the jammed cars released and soon production was back to normal. However, by the next summer the centre of the kiln had begun to sag. Investigations showed that the soil underneath was peat, which contained a

high proportion of flammable organic material. The red-hot concrete floor had set alight the peat, which had been quietly smouldering for a year. Fixing the problem was a huge job, recalled Clark: 'We started digging, tunnelling into the hot stuff, hosing it down, and bringing it out. We had the sprinklers going on it continuously. And we got it all out. We dug it right out, reinforced the kiln underneath . . . a little over 12 months and we had the whole thing fixed.' While the

repairs were being made the kiln was kept in full production. Enormously heavy, it was supported from underneath and raised centimetre by centimetre until it was level again. There were no commercially manufactured jacks big enough for this work, so the engineering shop made their own. The kiln was still in use when Crown Lynn closed 40 years later.

Jugs this shape were manufactured for many years in several different sizes. The jug in the foreground is made from the early yellow clay body. A design has been hand painted on top of the glaze by an unknown decorator.

TECHNICAL CERAMICS / During the five years of the Second World War, Ambrico's porcelain department prospered. While imports were restricted the factory made all sorts of precisely shaped dry-pressed porcelain items, usually for customers who had previously imported them. Postwar, the business, now managed by Geoff Amoore, had established a market niche and remained prosperous. Radiator bars and fuses for household fuse boxes were big sellers. Flat porcelain elements for the first electric stoves were also produced in large numbers. The elements had a spiral groove for the heating wires to be slotted into. During the war the neighbouring brick and pipe factory was also at full stretch, making sewer pipes for the American army bases in the Pacific Islands.

POSTWAR GROWTH / 'You might be interested to know that we operate two tunnel kilns. One 200 ft (60 metres) long gas fired and the other 100 ft (30 metres) oil fired. So although we live a long way from the centre of things we are not as backward as might be supposed.' Tom Clark in a letter to John Cowdery, 2 July 1946.

During the war years Tom Clark and his staff were working in isolation. They had done all they could to improve quality but there were still huge gaps in their technical knowledge and Clark was frustrated by the failures that occurred as a result. Already the factory employed about 400 people, and he felt there was a good future in domestic ware.

In 1946, not long after the war ended, Clark went to England in search of new equipment and new ideas. After touring the established English potteries such as Royal

Doulton and Johnson Bros, he felt that on the whole the team hadn't done too badly, but there was still ample room for improvement. He ordered English machines for preparing clay and for making cups and plates, and visited glaze makers Blythe Colour, who became long-term suppliers.

In Stoke-on-Trent Clark also met John Cowdery, an expert English mould maker from Royal Grafton pottery. Before Clark left on his overseas trip, Cowdery had come to his attention almost by accident. Back in a humdrum job after working for the British intelligence service in the Mediterranean during the war, he had written to New Zealand Insulators in Temuka looking for a new start. His letter was passed on to Tom Clark, who replied enthusiastically. In England Clark visited the Cowdery home, where John's mother was astonished that her tall visitor from New Zealand didn't wear a grass skirt. The two young men met after work and talked into the early hours.

Cowdery was sufficiently inspired by Clark's vision to pack up and move to the other side of the world, where he revolutionised mould making. Under his guidance the factory was able to make creditable slipcast jugs and vases, more elegant and finer than anything they had previously achieved. He also knew how to 'block and case' – to produce moulds that enabled the potters to slipcast hundreds of items that were exactly the same as the original.

At first the transition from England to New Zealand was not easy for the young immigrant. He was desperately lonely, homesick and ostracised by some workmates who resented having a know-it-all Pom tell

them what to do – even if he did know it all. However, a friendship with a young lady from the office, Bebe Woolright, soon cheered him up and in October 1948 they announced their engagement.

THE JEWELLERY / Around the mid-1940s an Auckland jeweller asked Tom Clark if he could make little ceramic 'buttons' or cabochons to set in silver bracelets and necklaces. Made of dry-pressed porcelain, the ceramic buttons were created in a range of round and oval shapes. Laboratory worker Mary Gardner painted them with flowing glazes developed by Denis McClure. After being fired they were ready for the jeweller to fit into mass-produced silver mounts. The quality was pleasing and the colours attractive and, said Clark, 'They sold like a bloody rocket.'

Melva Ockleston recalled that a considerable amount of jewellery was made, including bracelets and matching brooches. When a delegation of members of Parliament visited the factory, an attempt to lobby Mabel Howard with the aid of a bracelet almost ended in disaster. 'Tom had a bracelet made for her but she was a big lady and they had great difficulty doing it up; they had to enlarge it to make it go round her wrist.' It appears that there are very few of these interesting bracelets left – no doubt some lie forgotten in old jewellery boxes, their significance unrecognised.

Around this time the factory also made ceramic wreaths for decorating graves. Clark and his staff clearly remembered making the wreaths, which were circular with hand-applied ceramic flowers glazed

(Above) Bracelets made in the 1940s.

(Opposite page) A cigar jar from the 1940s or 1950s.

in pastel green, yellow and pink. However, examples that can be confidently attributed to Crown Lynn are very hard to come by.

THE CIGAR JARS AND SEPPELTS

BOTTLES / A simple cigar jar was one of John Cowdery's early moulding successes. Slipcast and glazed in a soft brown, it was made for an Australian manufacturer and importer of cigars.

Around this time the factory was also making lidded ginger jars and many thousands of slipcast liqueur bottles for the Australian firm Seppelts. Decorated in three colours, including a glistening greenish-brown flowing glaze, the bottles

were a technological and design benchmark. Colin Leitch, who joined the firm as a young accountant in 1946 and later rose to become general manager, recalled a disastrous fire which destroyed an early consignment of Seppelts bottles.

A CLAY BONANZA / In the immediate postwar years there was frenzied activity in the laboratory, which by this time was run by Denis McClure and 'Doc' Heine. New glazes were being developed, and the chemists were still trying to produce a white body, in both vitrified porcelain and earthenware. The laboratory staff were so caught up in the excitement that each week they gave up their Wednesday lunch hour to hold seminars on new products and processes. Night classes at Avondale College were also well attended. Chemists toiled long and hard to find a combination of clays which would come out of the kiln consistently white. Then, in 1948, the years of scouring the country for suitable materials finally hit pay dirt. Big deposits of white clay were discovered at Matauri Bay in Northland and at Mt Somers in Canterbury, and a mixture based on these two clays finally yielded the white body the chemists sought. For many years the company also took large amounts of clay from a deposit at Maungaparerua near Kerikeri.

GEARING UP FOR MASS PRODUCTION / Tom Clark came back from his first overseas trip bursting with new ideas and keen to redesign the factory for mass production. There was a ready market for everything he could make. After years of wartime austerity New Zealanders and Australians were

eagerly searching the shops for new dinner sets, mixing bowls, vases and other decorative items. Government departments needed dinnerware for their staff canteens, and the hospitals and New Zealand Railways wanted cups and saucers and plates. By 1948 factory reorganisation was in full swing. New machinery was installed and, after a building permit was 'at long last' granted, a casting shop was built at the rear of the new factory. The construction department was fully stretched with these projects, plus a number of smaller jobs including redecorating the cafeteria and building a first aid room. Probably the most important innovation was the first Prouty kiln – so named because it was designed by a Mr Prouty, an American. For the first time the

factory had a kiln that protected the ware from direct contact with smoke and flames caused by the burning oil used for heating. It was a continuous kiln, with ware stacked on tungsten carbide slabs moving slowly from one end to the other in a 12-hour cycle. Tom Clark bought the plans for the kiln from the United States and built it with fireproof bricks made at the Kamo brickworks. The new kiln greatly increased the range of decoration techniques and improved durability and finish.

THE ENGLISH DECORATORS / 'If it had roses and a bit of gold on it, at that stage of our history it would sell.' Tom Clark

After John Cowdery's arrival there was a huge improvement in the shapes being turned out, but decoration was still a weak point. Tom Clark and his chemists had worked out how to make coloured glazes, but they didn't know how to apply any other types of decoration. There was a huge gulf between the quality of English-made ware and the Ambrico products. Clearly the factory had to do better, and Tom Clark advertised in the Stoke-on-Trent newspapers for a decorating manager. Tom Bollington arrived in late 1947, and on his advice two English decorators, Doris Bird and Mary Baillie, were also hired. In English potteries hand decorators served a seven-year apprenticeship, so all three were experts at this type of work. Their skills, combined with the installation of the new Prouty kiln and a decorating kiln, revolutionised decorating.

Under Bollington's guidance the factory, for the first time, was able to use a range

36

This ivy leaf decoration was probably one of the first imported lithographs to be fired in the new Prouty kiln.

of techniques including applying lithographs or gold, hand brushing, lining and banding. 'Underglaze' decorations were protected with a top layer of clear glaze.

HAND DECORATING / Throughout the 1940s and 1950s decorations were applied by hand. Once the new kilns were in operation, lithographs imported from England could be used. This opened up a vast new array of decorations; the English had been making full-colour lithographs for decades and there were thousands of variations, mainly featuring flowers. Each lithograph was individually placed on the ware on top of the glaze, but they were properly fired and much more durable than the decorations done by Salisbury and Harwyn. All other decorations involved hand painting or hand stamping. Some early designs were decorated in a similar way to a child's colouring book. A design outlined in black was applied using simple transfers, then 'coloured in' by hand. Single colour transfers were used to produce monogrammed ware for hotels, restaurants, local hall societies and marae, initiating a market which proved highly profitable for decades.

Even simple lines around the outside of a plate or cup were hand brushed, with each item mounted on a spinning base before the lines were applied. In one day a hand painter could decorate about 60 pieces. Gold designs around the outside of a plate were stamped on – also by hand. Some of the hand-decorated items, especially in the early days, were signed on the base by the artist. For example, an ornate coffee pot in green and gold carries the words 'hand painted' and 'Doris Bird' as well as the Crown Lynn backstamp. Gold, which was either hand painted or stamped onto ware, required a great deal of skill to get right. It had to be fired at exactly the right temperature. If the kiln was too hot it vaporised; not hot enough and it would quickly wear off. The liquid gold colour was so valuable that the brushes used to apply it were sent away to a specialist factory which recovered the residual gold.

Doris Bird's husband Harry, also a 'pottery man', was employed to organise the warehousing systems. In keeping with the attitudes of the time the staff newsletter noted that 'Mr Harry Bird has joined us as

an expert in biscuitware warehousing' – but failed to mention Doris. For much of her 21 years at Crown Lynn Doris Bird headed the decorating team, working long hours when the factory was busy. Her daughter Gina remembered visiting the factory on her way home from school and walking past hot kilns and buzzing machinery to ask her mother what to cook for dinner. Her parents would often return to work after they had eaten. During the late 1940s seven English tradesmen moved their households to the other side of the world to work at Ambrico. Their skills were vital, but not all English immigrants stayed at the factory. Personality clashes and other frustrations drove some to find jobs elsewhere. A few, including decorator Mary Baillie and modeller Peter Cooke, moved to the rival Titian Studio, while others began completely new careers.

THE TOBY JUGS /
In the late 1940s Ambrico began making slipcast, hand-painted Toby jugs in imitation of Royal Doulton and other English manufacturers. The New Zealand versions appear amateurish in comparison: facial features are clumsily formed and the colours are garish. The Toby jug at left was also produced in plain green and brown. A less common shape shows a seated man with his mouth open, presumably laughing. A Santa jug was also made, along with a water jug depicting 'The McCallum'. Most examples are in the honey-coloured glaze but a few were painted in more lifelike colours.

HOPE FOR THE FUTURE /
By 1948 the early days of heavy, ungainly mugs were long since gone and the only major quality problem was that cup handles still tended to fall off. Tom Clark felt that his products compared well with ware from the English manufacturers Meakin and Johnson Bros. He didn't try to make fine bone china; the process was too complex and the machinery too expensive. He made durable vitrified hotel ware and relatively stylish, good-quality everyday domestic earthenware – dinner sets, jugs, mixing bowls, chamber pots, shaving mugs, vases and honey pots with lids. With the help of his English experts he had begun producing good-quality hand-decorated ware. The domestic market was solid, export prospects looked excellent, and there seemed no limit to how high the business could fly.

A SOCIAL SNAPSHOT /
Ambrico's social and sports clubs were very active in the 1940s. The October 1948 'Potters Pie' staff magazine gives an insight into the social activities. The 1948 annual social and dance in the Avondale town hall was attended by more than 300 people, including some of the 75 who had just been laid off due to the downturn in business. 'Mr Colin Leitch pleasantly and capably discharged the responsible duties of Master of Ceremonies.' Items included:

 'The Snake Charmers' by Kath Gillespie, Phyllis Clark and Nita Kahui.

 A Hula Girl's Dance by Mr Dave Garlick of the Kiwi Concert Party.

 A song – 'Why Am I Always The Bridesmaid' by Kath Gillespie, Phyllis Clark and Nita Kahui.

 Miss Pauline Sollett on the Piano Accordion played a well-known march, and 'Jealousy'.

The Ladies' Club held social evenings in the 'gaily decorated' Old Factory staff cafeteria. A bring-and-buy outside the New Factory raised £31 towards the £57.10/- cost of a piano. The Ladies hosted an evening for the Gentlemen's Club and a month later the Gentlemen reciprocated. On both occasions the assembled Ladies beat the Gentlemen hands down in a musical nursery-rhyme competition – this despite the fact that 'local booksellers have been surprised at the sudden demand for books on Nursery Rhymes, and Ambrico gentlemen are to be discerned in odd corners at odd hours memorising rhymes'.

The Gentlemen's Club met on the first Friday of every month. 'All the meetings have been very well attended and enjoyed so much that the stewards of the club have difficulty in bringing the proceedings to a close at the stipulated time.'

In an away game in Tauranga the Football Club was trounced 18–5 by the Rangatau Club. The opposing team 'was very heavy and fast and at times it showed brilliance, whilst our boys were fatigued from the long trip the day before'. Any hurt pride was restored at the after-match function where 'so successfully did the two crowds mingle that honours were conferred on some of our members. Arthur Godfrey was elected Rangitira to the Mangatapu Pa with great hilarity, and the Elder of the Pa donated to the Team Manager, a 9-acre farmlet.' On the trip home the team stopped at the hot pools in Matamata, where 'Messrs Stevenson and Godfrey astounded all spectators with their exhibition of flat diving'.

50s

THE WILDERNESS YEARS

1949–1959

By the beginning of 1948 Ambrico was the largest pottery in the Southern Hemisphere. Its 300 workers produced six million pieces a year, half of which were exported to Australia. Then, in August 1948, disaster struck. Peter Fraser's Labour Government revalued the New Zealand currency and Ambrico lost its 20 per cent price advantage in Australia. Overnight most Australian orders were cancelled and Ambrico lost half its market. At the same time the company began to meet fierce price competition in New Zealand, mainly from duty-free crockery imported from Britain as wartime restrictions eased. Orders fell away and the workforce was pruned to 100. Through the 1950s the factory was buffeted by a series of economic events outside its control, including altered import regimes and changes to the value of the New Zealand currency. The company struggled to survive, and Tom Clark often referred to this period as the 'ten years in the wilderness'. Although he never seriously entertained the idea that the business would fold, Clark found himself under siege from creditors.

With the Australian market gone, Clark began urgently seeking new outlets in Canada and the United States. He also diversified,

[Opposite page] The Crown Lynn swan was manufactured from the 1950s to the 1970s.

1950s

After the war the domestic refrigerator became an accepted household item. The New Zealand manufacturer McAlpine commissioned Crown Lynn to make a water jug (opposite page) which would fit in its fridges. Over the years the factory produced a range of jugs (below) for various companies.

making new products including salad ware, cake plates and souvenir ware. There was even an attempt to manufacture ovenware, though the mottled brown casserole dish was not a huge commercial success.

CROWN LYNN IS BORN / In 1948, in the midst of financial difficulties, Tom Clark decided that his business should have a new name – Crown Lynn. He chose 'Crown'

because of its connotations of quality and its connections with England, and 'Lynn' after New Lynn, the suburb where the factory was based. At around the same time the porcelain department was renamed Crown Lynn Technical Ceramics and became a stand-alone business managed by Tom Clark's brother Malcolm. At first, Crown Lynn products were identified by stick-on labels rather than permanent backstamps.

THE DESIGN DEPARTMENT / From the very beginning Tom Clark recognised the importance of good design. In those early days, though, design possibilities were limited because there were so many technical challenges to overcome. Clark and his team did their best to create pleasing shapes for the original pre-war mugs and hot-water bottles, but the first priority was to make the things retain their shape in the kilns.

The need for good design continued to haunt Clark and around 1945 he had asked John Weeks, a leading artist at the Elam School of Art, to design a dinner set. Sadly, the designs he created could not be reproduced with the factory's primitive machinery. During his second overseas trip in 1948, Clark was captivated by ware from Arabia Potteries in Finland and Sweden's Upsala Ekeby – everything they produced screamed 'style' and 'design'. He was inspired to put more effort into designing his product back home. 'Basically I was a technologist and an engineer. I thought that if I was going to do anything with design I would need somebody I could depend on.'

So Tom Clark asked David Jenkin, a young Elam graduate whom he had hired in 1945, to head a new design department. That was the beginning of a hugely productive partnership that lasted more than 30 years. Jenkin's quiet, low-key approach fitted well with the Crown Lynn culture. He was able to promote good design throughout Crown Lynn as well as work with the powerful and sometimes overbearing Tom Clark. In 1951 the equally unassuming Tam Mitchell was hired, and quickly promoted to chief modeller, creating one-off items from which moulds could be made. Summer or winter, he worked in bare feet, shorts and a shirt.

But he was an exceptional modeller and stayed with Crown Lynn for decades. Tom Clark remembered Mitchell's productive partnership with David Jenkin: 'Dave would do the design on a piece of paper with a charcoal pencil and that was all we needed, because of this wonderful guy Tam. Tam would make the model and then the mould makers would make the masters from that and away we'd go.'

WHARETANA WARE / In a bid to break into the souvenir market, in the late 1940s and

The Wharetana Ware range included several different types of wall plaques.

early 1950s Crown Lynn began manufacturing Maori-influenced Wharetana Ware. Modeller Harry Hargreaves used a new technique to create the new range. One-off clay models were made, and at a certain stage of dryness Hargreaves used a sharp knife to incise crisp geometric designs into the leather-hard clay. These originals were then used to make slipcasting moulds for mass production. The line included trinket boxes, wall plaques, ashtrays, a Maori canoe and other designs. They were glazed in gloss brown and usually oversprayed with various greens. Often a paper label explaining the

origins of the Maori design was stuck to the base. Wharetana Ware was not a great success and was discontinued around 1952.

FANCY FAYRE AND WENTWORTH WARE /
After the sudden slump in the Australian and domestic markets, Tom Clark was constantly searching for new items to sell. He began making imitations of popular English styles. One of his early attempts was the slipcast Fancy Fayre range, modelled on salad leaves in imitation of English Carlton Ware. Fancy Fayre Salad Ware was glazed in green, red or yellow, with tomatoes hand painted in red.

The range included jugs, plates and dishes, a condiment set and a salad bowl with drainage holes and tomato-shaped feet. Yet another imitation of English ware was the Wentworth Ware range, manufactured in the mid-1950s and early 1960s. Dishes and plates in various sizes were decorated in relief with flowers including the New Zealand native clematis. Most were marked underneath but some also carried a Wentworth Ware sticker. They were glazed mainly in pastel colours of yellow, pink and green, but some bolder red dishes were also produced. Most were edged in gold. Some slipcast vases, finished in green, were also marketed under the Wentworth Ware brand.

THE FIRST CAKE PLATES / Under the leadership of the new decorating manager,

Fancy Fayre (right) and Wentworth Ware.

Tom Bollington, the factory was also able to produce credible imitations of English cake plates, cups and saucers, and coffee pots. Some were decorated with airbrushed glazes, others with hand-applied floral transfers. Almost without exception they were trimmed in gold. By 1954 the factory was also making lidded dishes for serving vegetables.

THE TEAPOTS / Although Crown Lynn made countless jugs in various shapes and sizes, few teapots were manufactured. For a tea-drinking nation, in the days when loose-leaf tea was the norm, it is surprising that Crown Lynn did not fill this gap in the market.

Because of their irregular shape teapots are awkward to make, but by the late 1940s Crown Lynn certainly had the ability to make them. Throughout the company's history teapots were produced but never in large numbers.

THE TERRACOTTA WARE / During the early 1950s Crown Lynn produced a small amount of terracotta ware. It was not made in large amounts because of the fear that the reddish-brown clay would stain the machinery – when every effort was being made to produce a white body. The terracotta items were all hand potted, thus avoiding contamination of the machinery.

(Below) This teapot with its distinctive recessed lid was made in the late 1950s or early 1960s.

(Opposite page) Cake plates were made in a wide variety of colours and decorations. The plate on the right is backstamped 'Regal Potteries', the others 'Ascot'.

ERNIE SHUFFLEBOTTOM / On a postwar visit to England, Tom Clark was struck by the luminous white vases being created by Keith Murray at Wedgwood. Murray used an unusual technique. He formed his pots on a potter's wheel, then dried them to a leather-hard consistency before hand turning them on a lathe to create crisp shapes and sharp edges. A soft, almost pearly white glaze offset the turned vases to perfection. Seeing an opportunity to make a more upmarket product, Clark advertised in the Stoke-on-Trent newspapers for a thrower and turner. He was delighted when Ernest (Ernie) Shufflebottom applied and promptly hired him. Shufflebottom was a protégé of Keith Murray and brought with him a selection of excellent designs – which Clark later realised actually belonged to Murray. However, Shufflebottom soon redeemed himself by creating his own designs. Meanwhile, the hunt was on to create a soft white glaze. Although matt white glazes were probably made before this time, perfection was not achieved until 1948 when the new Prouty kiln made it possible to fire glazes which, said the staff newsletter, had 'a nice texture reminiscent of the bird's eggs of one's youthful pilfering'.

48

Ernie Shufflebottom's arrival in New Zealand in 1948 coincided with an upsurge in the domestic art of flower arranging and his handsome clean-lined pots sold like wildfire. Big white pots ablaze with gladioli stood staunch on either side of church altars up and down the country, while smaller vases adorned thousands of mantelpieces. Housewives attended flower-arranging classes or consulted Constance Spry's books for inspiration. Spotting an excellent marketing ploy, Crown Lynn sponsored flower-arranging competitions. For years there was a ready market for Shufflebottom's work and he and his assistant Jack Aberly turned out rafts of hand-potted vases, candle holders and lamp bases. Most of Shufflebottom's pots were glazed in white, but some were also finished in pastel colours. At first Shufflebottom's work was in keen demand, but an economic

It is likely that some of the vases marked 'hand potted' were made by Ernie Shufflebottom's assistant Jack Aberly. It is also probable that the work of other less high-profile potters who worked at Crown Lynn bears the 'hand potted' inscription.

(Opposite page) This vase was hand potted by Daniel Steenstra. The decorator is unknown. Most highly individualistic items like this vase were 'homers' – they were never intended to be sold commercially and were taken home for personal use.

(Below) Coronation mugs from 1953.

downturn and changing customer tastes meant that by the mid-1950s Crown Lynn faced a growing stock of unsold white pots. Around 1956 Ernie Shufflebottom was 'let go' and returned to England to take up his old job at Wedgwood.

THE COMMEMORATIVE MUGS / In the 1950s Crown Lynn made a series of commemorative mugs. The first was a souvenir mug made for the 1950 Empire Games in Auckland. The slipcast mug was designed by David Jenkin and modelled by Ernie Shufflebottom. The English-made decorative transfers and gold lines were applied by Doris Bird. Ten

thousand were sold to games supporters from New Zealand and overseas. In the same year a similar mug was created to mark the centennial of the province of Canterbury.

After the success of the Empire Games souvenir, in 1953 Crown Lynn created another limited edition mug, this time to commemorate the coronation of Queen Elizabeth II. This was also a team effort. It was designed by David Jenkin and the English-trained modeller Peter Cooke, hand thrown by Ernie Shufflebottom and decorated by Doris Bird. Only 1000 mugs were produced. At the same time Crown Lynn also made a much simpler – and no

doubt less costly – version of the coronation mug. Glazed in pale yellow, the slipcast mug was embossed with a portrait of the young Queen.

DANIEL STEENSTRA / Daniel Steenstra started work at Crown Lynn in 1953. Newly arrived from Holland, he quickly made himself a niche, creating hand-potted vases, ashtrays, candle holders and jardinière. Tom Clark later recalled that Steenstra was the 'sharpest, smartest thrower you ever saw in your life', quickly turning out hundreds of very similar items. He could create a pot then make a lid that fitted exactly – without measuring anything. Steenstra decorated some of his products with fluting and other textures, using small tools to pare away surplus clay while the piece revolved on the wheel. He also sometimes decorated his hand-turned items with coloured slip. Generally, Steenstra was not a mass-production hand painter, though collegues say he did sometimes paint his products. Most of his works were glazed in a single colour or decorated by Frank Carpay, Doris Bird, Eileen Machin and probably other hand painters. Part of Steenstra's job was to assist in the development of forms which were later mass-produced. He was able to create a three-dimensional form from Dave Jenkins' drawings, and to change the shape on the wheel until both men were satisfied with the result. It is likely that he also produced some vases in the style of Ernie Shufflebottom.

By the mid-1950s many New Zealanders were beginning to weary of the sameness of mass-produced moulded china and, as the *Auckland Star* put it, 'No beautiful home is

51

complete without some examples of the potter's handicraft.' Daniel Steenstra's demonstrations on the potter's wheel at trade fairs and industrial shows attracted fascinated crowds, many of whom had never seen a potter at work before. In 1960 newspapers enthusiastically reported that Steenstra had shown Mrs Margaret Young, Taranaki's 'Most Gracious Mother', how to create a vase on his potter's wheel. Mrs Young, who won her title against competition from 100 Taranaki mothers, told reporters that she could see herself taking up potting as a hobby. Steenstra stayed with Crown Lynn for nearly two decades, although during downturns he sometimes had to take temporary work at Crum's brickworks up the road. Through the 1950s his hand potting was in demand, but in the 1960s as Crown Lynn geared up for mass production his skills became redundant. After a frustrating period working at production jobs, he moved on to Beach Artware, a smaller craft pottery. He left behind a legacy of elegantly shaped vases, salt and pepper sets, candle holders and other items – and took with him his wife Wendy, whom he had met in the accounts department. Daniel and his brother Thijs, an expert on glazes, started at Crown Lynn at the same time. Thijs never potted for Crown Lynn, instead working at various jobs around the factory.

MIREK SMISEK / In 1950 young Mirek Smisek arrived on Crown Lynn's doorstep looking for a job. He began in the clay preparation department and was soon promoted to assistant to Ernie Shufflebottom. After practising in his lunch hours, Smisek became proficient at making vases on the potter's wheel and was given his own workbench. His best-known Crown Lynn work is the Bohemia Ware range, named after his birthplace. These vases are decorated in a style known as sgraffito. Hand-potted vases were dipped in a brownish semi-gloss glaze, then Smisek used a thin steel pencil-like tool to trace patterns through the glaze, exposing the pale body beneath. Most of these vases are around the same height – about 12 centimetres. Smisek had a yen for independence and after only 18 months he left Crown Lynn. Before long he set up his own hand-potting business and later made the pottery for the *Lord of the Rings* movie trilogy.

FRANK CARPAY / Frank Carpay began working for Crown Lynn in 1953. He was recommended by head designer David Jenkin and, after a chat with the thin intense Dutchman, Tom Clark hired him on the spot. At first Carpay hand-brushed decorations onto standard Crown Lynn shapes including meat platters, hotel milk jugs, pie dishes,

dessert bowls and plates. He soon became frustrated with the limitations of these forms and designed his own platters, jugs, lamp bases and vases. Carpay was not a potter so his designs were executed by Crown Lynn's hand potters. Handwerk was the end result, a highly distinctive range, harmoniously shaped and painted in dramatic brush strokes. In his first six months at Crown Lynn he produced 175 new designs. Carpay had spent some time with Picasso in Europe and was strongly influenced by his bold style.

Years later Clark remembered watching Carpay decorate his spectacular platters: 'I had that same feeling about it as I had about Ernie Shufflebottom … it was beautiful, beautiful, beautiful. But the big problem was to get him to appreciate just how much ground I could afford to give him. What he made had to be sold.' Clark and David Jenkin promoted their new designer with exhibitions, newspaper articles and public demonstrations of his hand-painting techniques. Sadly, although there was considerable coverage in the local media and respect from design critics, the 'average New Zealander' was simply not interested. In retrospect Clark believed that the work was too unconventional for the tastes of that time. Most English potteries were still turning out delicate fine china adorned with roses and violets, and the buyers simply couldn't accept Carpay's bold,

53

(Opposite page) In early 2005, only a few months before Tom Clark's death, an Auckland auction house sold off a large private collection of Crown Lynn. The sale included a spectacular Frank Carpay platter which Clark was determined to own. As the auctioneer dropped his hammer he called out, 'Sold – to that gentleman over there.' He had no idea that the successful bidder had once been Frank Carpay's employer.

(Left) A selection of hotel jugs. The pattern at far left was made especially for the Intercontinental Hotel.

54   chunky Continental style – much less his oil bottles decorated with images of semi-naked women. 'When I think about the people we used to try and sell to, people like Smith and Caughey's, you would go and talk to the buyer and he'd see this stuff and you would see them shuddering, because it wasn't British.' Frank Carpay, though, was not of a mind to change his designs to meet the market. In Europe his designs were accepted and he was determined to wage a 'war against the rosebuds' in his new country. Unfortunately, most New Zealand householders still preferred rosebuds. To compound his problems, cheaper imported ware in a similar style was beginning to appear in New Zealand shops.

Carpay, said Clark, was furious that his work didn't sell. He was argumentative and difficult – but without doubt the most exciting and interesting designer who had ever worked at Crown Lynn. Clark was surprised and disappointed at the lack of buyer interest, but he had to take a pragmatic approach. As unsold product stacked up in the warehouse, he decided he could no longer afford to keep Carpay. After only three years, to Clark's enduring regret, he was sent down the road. 'To me that was really sad. He deserved a better chance than we were able to give him.' Connie Clark, wife of Tom's brother Malcolm, was an early admirer of Carpay's work. She bought several pieces from the pottery, and remembers picking up many more for a few shillings apiece in second-hand shops around Auckland.

Fifty years on, Carpay's Crown Lynn work still looks fresh and modern, and Clark believed it would still stand out in another 50 years. 'It's the one thing about the whole of Crown Lynn that's outstanding … it was just magic, just magic.'

THE HOTEL JUGS / One of Crown Lynn's most enduring designs was the simple straight-sided jug – shape number 715. Slipcast in at least six sizes, it first made an appearance in the late 1940s or early 1950s and continued in production almost until Crown Lynn closed down. Popular with institutions as well as ordinary households, the jugs dispensed milk or lumpy custard to boarding school students and hospital patients throughout New Zealand. Most jugs were white, but some were banded around the top in red, decorated with simple patterns or glazed in plain colours. A few were decorated by Frank Carpay. Glazed in rich maroon with the railways logo around the

top, the jugs were included in ware made for New Zealand Railways. In 2005 a pair of lime green examples were still doing duty in the visitor's kitchen at Greymouth Hospital.

THE WHITE SWANS / 'It sat in the picture window facing the street, with the white venetians poised above its head, and terylene drapes artfully criss-crossed on either side. Inside its hollow back sailed a bunch of red plastic roses, fading to fleshy pink in the sun.' Rosemary McLeod, *New Zealand Listener*, 30 June 1984.

In the 1950s and 1960s it seemed that every second New Zealand house had a Crown Lynn swan. The original shape was copied by David Jenkin from an example brought back from overseas by Tom Clark. And thus began a decades-long success story. Some of the very early swans bear the old Ambrico fractionated marks. They were decorated in pastel colours and trickle glazes or in a glossy white finish. From the late 1940s mass-produced swans were glazed in the soft white which was used on Ernie Shufflebottom's work. A few sported gleaming yellow beaks; others were glazed in dramatic black. For decades Crown Lynn turned out slipcast swans in their hundreds. As late as 1973 a white swan still featured in a Crown Lynn newspaper advertisement along with an array of matt white vases.

The swans made excellent gifts. Frank Fitzpatrick, who worked in the Crown Lynn laboratory, gave a swan – the biggest in the range – to each of his friends on their wedding day. In later years many swans ended up as forgotten and grubby repositories for plastic flowers, old pens

and drawing pins. During the flower power revolution of the 1960s many young people saw the Crown Lynn swan as a symbol of all that was wrong with their parents' concept of design. Nevertheless, 15 years later they were enjoying a revival of sorts. In 1984 the influential columnist Rosemary McLeod wrote in the *New Zealand Listener* that the swans could be bought at auction for as little as $10 a pair, and predicted that they were on the brink of a comeback.

MATT WHITEWARE / From the late 1940s to the 1970s swans were not the only white vases made by Crown Lynn. The factory also turned out thousands of slipcast vases, most finished in the same white glaze as Ernie Shufflebottom's vases. Mass-produced in moulds, this range does not have the exclusivity of Shufflebottom's hand-potted

work. Much of the whiteware was sold under the Flair Art Pottery backstamp. Some was unmarked, or had various stickers attached. The plain whiteware was extremely popular as vases or mantelpiece decorations right through the 1950s and into the 1960s. Over the years the company made a huge variety of items, including shallow bulb bowls, flower pots, a hand-made water jug, mugs 'for the home brew expert' and novelty vases in the shape of animals including a horse, dog and deer. There were 17 different wall vases. In 1963 there was a continually changing range of over 100 shapes in production. Some shapes were available in matt black or pastel colours as well as white. Crown Lynn promoted the white vases as the perfect Christmas gift, advising retailers to brighten up a display with a bunch of flowers – real or artificial: 'Chances are the customer might wish to

(Above) A popular whiteware pattern.

(Opposite page) In 1959 an Auckland flower-arranging competition that was sponsored by Crown Lynn had almost 200 entries. The grand prize of an Ultimate radiogram was won by Mrs H. L. Beatty of Silverdale for her arrangement of lilies and Prince of Wales feathers designed to create the impression of movement and water.

buy the complete arrangement if you use artificial flowers in the display.' Crown Lynn also produced a range of lamp bases, many in the same glaze as the vases. Some were hand turned, others slipcast. Although most were white, some black examples were also made.

THE EGG CUPS / For decades thousands of young New Zealanders woke up on Easter Sunday to the sight of a Crown Lynn egg cup, usually complete with a chocolate marshmallow Easter egg and a tiny fluffy toy chicken. The first egg cups from the 1940s were relatively primitive, but later examples were more carefully modelled. Some were made for one-off promotions, others for special customers. The early examples were glazed in a single colour, but later decorations were more complex. Crown Lynn egg cups are now enthusiastically collected, but they can be difficult to identify because not all are marked on the base. Although they were good sellers, the china egg cups were never especially profitable because they were sold cheaply and their market was limited to one or two weeks a year. The egg cups were all slipcast, many by Rod Hendry, whose speed became legendary. He would lay out rafts of plaster of Paris moulds – each making six egg cups – on his workbench. In a smooth continuous process, he would work around the room, filling moulds with liquid clay, leaving them for a few minutes then emptying them as they set.

THE BOWLS / Right back to the Ambrico days, Crown Lynn produced mixing bowls. Early examples were whitish, then some were decorated with a single colour glaze. Later, during the late 1950s, Crown Lynn produced hand-painted Fiesta Ware bowls decorated with bright bands of colour. The Fiesta Ware range also included a dinner set, bowls and a condiment set. Since the bands of colour were applied over the top of the glaze, many examples are damaged. Later still the pottery produced bowls decorated with underglaze colours.

THE SOCIAL CLUB GROWS / Through the 1950s the Crown Lynn social club found its feet. There were drinks and sometimes ballroom dancing on Friday night, and dances and parties throughout the year. The most keenly anticipated was the annual Christmas party, where Santa gave out presents to the children. As a child, Doris and Harry Bird's daughter Gina remembers singing 'Buttons and Bows' at one of these events, standing up on stage in her home-made dress made of yellow taffeta overlaid with green net. For a fancy dress ball Doris and Harry turned up in Mickey and Minnie Mouse costumes which they had spent hours making. They were definitely the stars at the ball, but there were no prizes, recalled Gina. 'They told

59

(Above) A selection of Crown Lynn egg cups.

(Opposite page) These vases are probably from the 1960s.

Mum and Dad they wouldn't be considered because the costumes had to be home-made and not hired. They didn't believe Mum when she said that she and Dad had made them. She was so upset!'

The social club also supported sports teams including rugby and netball, and women's marching, which was then in its heyday. There were senior and junior marching teams, which were successful in local competitions and marched in the Farmers Christmas parade. For a time in the late 1950s and early 1960s the team was led by Elaine Miscall, who won the Miss New Zealand title in 1963.

THE FACTORY IN THE 1950S / Despite the economic downturn of the 1950s, Crown Lynn was busy and productive, employing between 100 and 150 workers. Mary Stewart, who operated a 'cranky old adding machine' in the accounts department in 1950, had abiding memories of primitive equipment, noise and dust. Much of the work was repetitive. Women sat all day, every day, sticking handles on cups. Tom Hodgson, who worked for Crown Lynn on and off for 30 years, remembered his first job as a schoolboy, working on the 1955 Christmas refit. 'The factory was old, even then. It was made with second-hand timber and whatever. It was like a cavern with many nooks and crannies. It was an absolute wonderland for someone 15 years old, believe you me. There were tanks in the ground and tanks above the ground and belts and pulleys – all dangerous stuff. It's a wonder no one was maimed or killed. But no one was.'

THE 'BRITISH' BACKSTAMPS / In the earliest days, before he had named his company

Crown Lynn, Tom Clark proudly stamped his product 'Made in NZ' or 'Made in New Zealand'. Later, probably in the early 1950s, he felt that his product would sell better if buyers believed it was made in England, the traditional home of quality china. At first he simply failed to mention where an item was made. Backstamps with English-sounding names like Fancy Fayre Salad Ware, Ascot and Wentworth Ware began to appear. Later, probably from the mid-1950s until the mid-1960s, Clark brazenly labelled his New Zealand-made products 'British'. The series included Ferndale British, Kelston Ware British, Gigi British, even Covent Garden British. Understandably, this ploy infuriated both the English potteries and the importers of English-made ware, but Clark justified his actions on the grounds that, like all New

(Opposite page) Crown Lynn bowls. Backstamped Crown Lynn Star and Tiki (left), Fiesta Ware (rear), Aero British (front) and unmarked (right).

(Below) A night out in the early 1960s. From left: Sylvia Sawyer (front), Doris Bird, Harry Bird, name unknown, Fred Hoffman, Bebe Cowdery, John Cowdery, Henry Sawyer. At rear: Colin Leitch, Isabel Leitch, Joe Elliott.

61

(Above) Applying cup handles was a tedious and painstaking job. This photograph was probably taken in the 1950s.

(Opposite page) These cup and saucer sets carry a diverse array of backstamps. Clockwise from bottom left: Golden Bouquet, Crown Lynn, Ascot, Lido British, Sylvia Rose and Crown Lynn. Centre: Bouquet.

Zealand citizens at this time, he carried a British passport. 'The government gave me a wigging. I pulled out my passport and I said the day you give me a New Zealand passport that's the day I'll put New Zealand on the back of the product.' Eventually, he was forced to capitulate and put the true country of origin on his products. Interestingly, both versions of some backstamps exist – for example Gay Gold British appears, as does a simple Gay Gold. On close inspection it appears that the offending word has simply been trimmed off the stamp.

THE WAYWARD CUP HANDLES / Through the 1950s Crown Lynn cups were famous for their unreliable handles – tip your tea dregs out the back door and the cup would fly out into the garden, leaving you holding yet another handle without the cup. The problem

was finally solved in the late 1950s when engineers discovered that a specially mounted Schick razor blade would cut a curve in the end of the handle that exactly matched the curve of the cup. Provided both handle and cup had been dried to exactly the right stage, and provided the liquid clay slip used to stick handle to cup was exactly the right consistency, the handle bonded securely to the cup. Tom Hodgson recalled that the good news went round the factory like wildfire and the men who solved the problem achieved hero status. After this breakthrough Crown Lynn's quality control staff had to test the strength of cup handles. They used a small calibrated pendulum that swung back and forth against the handle with steadily increasing pressure. More often than not the cup broke before the handle came loose. By the 1960s public suspicion had abated. In 1968, under the heading 'Unsmash', the *Wanganui Chronicle* trumpeted the virtues of Crown Lynn when a cup survived undamaged after falling three storeys from a windowsill onto the footpath.

THE FIRES / Over the years Crown Lynn suffered several devastating fires. There were no fatalities, but much of the early history of the pottery was lost. In 1945 the laboratory was badly damaged, and in 1956 a much larger blaze almost burnt down the entire factory. These fires destroyed almost all evidence of the early days including the original one-off experimental pieces and old notebooks and textbooks. In the 1956 blaze piles of banknotes held in the safe were charred almost beyond recognition, but Reserve Bank staff from Wellington

During the 1959 factory tour, Member of Parliament Mabel Howard meets head designer David Jenkin.

64 painstakingly peeled apart the scorched layers one by one with tweezers. The fire was a major event in the small community of New Lynn. Firefighters were hampered by a lack of water pressure due to 'thousands of people watering their gardens'. Police fought to keep back the crowds as there were fears that an underground tank containing 18,000 litres of fuel could explode. Clay worker Jim Stickings was the first to notice the fire and tried to put it out. He told reporters that as the flames spread he remembered the stored oil. 'The first thing I thought of was there's 5000 gallons of oil stored about 20 yards from here. I'm off.'

During the following decades there were many more fires at Crown Lynn, some large, some not so large. Sometimes plumes of smoke could be seen right across West Auckland. One fire cut short Alan Topham's

nice lunch with Government Stores Board representatives. Instead, he spent the afternoon making sure that everyone was safe and arranging for the mess to be cleaned up. The huge timber columns supporting the roof of the drying room charred but didn't fall, while the steel columns in the new decorating room folded like melting butter. Another blaze sent crockery raining down from the collapsing top storey onto firemen working below.

A NEW OPTIMISM / Throughout most of the 1950s Crown Lynn had struggled to survive. Frustrated, Tom Clark took to car racing and in November 1957 nearly died in a crash at Bathurst in Australia. Still in hospital the following January, he heard heartening news. A foreign exchange crisis forced the Walter Nash Government to apply temporary import

restrictions and, for the first time since the war, supplies of imported goods were severely limited. There was no option but to buy New Zealand made, which was excellent news for Crown Lynn.

In a new mood of optimism the company cranked up production. The existing workforce found fresh energy and droves of new workers were hired. Advertisements offered 'ladies and girls' £8.10/- a week with the possibility of overtime work.' In only 12 months the factory doubled its output and the number of designs quadrupled. Management struggled to maintain quality with so many new and untrained staff. In 1959 Crown Lynn made eight million items and supplied 40 per cent of New Zealand's tableware market. At this time the whole of New Zealand was undergoing a period of intense growth. In only two and a half years, 48 new factories and 26 new commercial buildings had gone up in the suburb of New Lynn.

It soon became obvious that Crown Lynn needed to mechanise to meet ever-increasing demand. The decorating department was the first to be upgraded. A conveyor belt eliminated unnecessary handling – gone were the days when a hand decorator prided herself on the number of cups and saucers she could carry balanced on her outstretched arms. The first automatic Murray Curvex machine arrived in 1959 and after a frustrating 18 months of modifications it greatly speeded up decoration. Previously, most decorating was done manually, with rows of women either hand painting or applying transfers one by one, but the Murray Curvex could print directly onto flatware. At first the machine was limited to one-colour prints, but it covered the entire surface of a plate and it could print concave surfaces such as the inside of a soup bowl. Best of all it was fast. With only one operator the machine could print up to 3500 pieces in an eight-hour day. This was a huge advance on previous decorating methods, and over the next decade or so Crown Lynn installed two more.

Crown Lynn's engineers developed new machines to spray coloured glazes and to stamp gold patterns around the edges of cups and saucers. A new jiggering machine made it possible to mass-produce cups at the rate of 3500 a day. Quality control, too, was boosted; products were tested for their resistance to breakage, chipping and crazing, and for their capacity to withstand sudden changes in temperature.

The search for new and faster machinery extended to the United States and Europe. In 1959 factory manager Fred Hoffman and production manager Colin Leitch spent 10 weeks overseas, investigating the latest manufacturing methods and, according to a local newspaper, 'assessing what trends will be important not only in a year or two, but in five and ten years' time'. They visited potteries and suppliers, and took a tape recorder to register impressions.

THE 'SNOB' SCANDAL / From the mid-1950s, as the quality of his product improved, Tom Clark found himself battling against entrenched prejudices. Many householders still spurned New Zealand-made Crown Lynn in favour of 'good' china from England; Alfred Meakin or Johnson Bros at the very least, Royal Doulton or Wedgwood if you could afford it. Crown Lynn was seen as second best, still struggling to overcome its reputation as a manufacturer of thick, heavy cups whose handles fell off. Eventually, Clark lost his patience and complained publicly that New Zealanders were snobs. This comment was reported in the newspapers and created a small furore. By this time Crown Lynn's product was as good as the everyday china coming out of England, but that made little difference to a buying public who still referred to Britain as 'home'. Attitudes slowly changed, though, and through the 1960s New Zealanders began to feel pride in their local factory.

THE PRIME MINISTER'S LOVING CUP / By the end of the 1950s there was no doubt that Crown Lynn could produce quality goods. On 18 July 1959 the company presented Prime Minister Walter Nash with the one hundred millionth item made by Crown Lynn. The elaborate loving cup, so named because it had a handle on either side and could thus be used by two people at once, was finished in white and burnished gold. The cup was designed by David Jenkin and decorated by Doris Bird – described in the newspapers as an 'Auckland housewife'. On the same day as the presentation, a large contingent of members of Parliament and industry leaders toured the factory. The creation of the cup received an enormous amount of newspaper publicity. Doris Bird said that making the intricate object was a real challenge, and at least six attempts were made before a perfect cup was produced.

65

60s

FULL STEAM AHEAD
1960–1969

During the 'wilderness years' of the 1950s Tom Clark had sworn he would build his company into something to be reckoned with – big enough to make the powers that be sit up and listen. Through the 1960s Crown Lynn continued to expand. New printing machines were installed, and a new laboratory and more warehouses were built. By March 1963 there were 360 staff, turning out eight million pieces of crockery a year. Crown Lynn supplied half the total New Zealand market for domestic china, and its warehouses held up to one million finished pieces. The factory made 80 different dinner sets and a huge range of cups, saucers, plates, jugs, ovenware, ornaments, bowls, nursery ware, lamps, vases, and vitrified tableware for hotels, hospitals and restaurants. The factory was huge and noisy. Hissing air valves from the Murray Curvex printing machines, clanking crockery in the grading department, rumbling clay preparation machines – all gave Crown Lynn an air of action and urgency.

The ceramics division, manufacturing insulators and electrical fittings, was also in good heart. In this field there was competition from the South Island factories Neeco and Temuka Potteries, but in an expanding economy there was room for all three to make a living. Importantly, the attitudes of the New Zealand public had begun to change. Tom Clark felt that he had at last begun to overcome the prejudice which led householders to buy English tableware in

(Opposite page) Crown Lynn produced a number of hand-painted designs, but Fleurette was by far the most popular.

preference to New Zealand made. He said that he was not trying to produce the fine china associated with names like Wedgwood and Doulton, but his everyday medium-price tableware was as durable and looked as good as similar products from Johnson Bros and Meakin. In fact, said Clark, 99 out of 100 customers could only tell the difference by looking at the trademark stamped on the base.

THE REPLACEMENT POLICY / In the late 1950s and through the 1960s Tom Clark and his managers hit upon a hugely successful marketing idea. They selected five attractive middle-of-the-road patterns and guaranteed that their customers would be able to buy replacement pieces for any that had been broken. They chose five patterns that they thought would sell – Autumn Splendour, Golden Fall, Shasta Daisy, Green Bamboo and Fashion Rose.

Colin Leitch and Fred Hoffman had discovered an almost endless supply of cheap lithographs from a New York manufacturer, so Crown Lynn was able to assure New Zealanders that the patterns would be available for at least five years. The patterns could be bought as complete dinner sets or piece by piece from racks in the big department stores. Cups and saucers and plates were the big sellers but vegetable dishes and other extras were also on offer.

Alan Topham, who was marketing manager at Crown Lynn at the time, said that the 'gutsy' introduction of the replacement policy created enormous excitement. No supplier had ever made that commitment before. 'They weren't the most magnificent designs in the world; in fact, Shasta Daisy you couldn't even call a design. It was just a little group of flowers, but it had broad appeal, and broad appeal meant that it sold well – and it had the added replacement connotation.'

68

From left: Golden Fall, Fashion Rose and Shasta Daisy.

These five unassuming patterns occupied a very profitable place in the market for over ten years. From 1960 until at least the end of the decade, Autumn Splendour was Crown Lynn's top seller. It remained popular well into the 1970s and was still being sold in 1977. Many young couples took advantage of the replacement policy. Now in her fifties, Christchurch-based Sandy Donohue initially bought an Autumn Splendour dinner set when she was first married in 1970. 'It was sold on big racks and you could buy whatever pieces or numbers that you wanted. Then my marriage broke up and I never saw my china again. So I decided a couple of years

ago to collect Autumn Splendour once more. I have had so much fun hunting round in second-hand shops finding it.' These days Donohue is again using her Autumn Splendour dinner set, though now it is kept for special occasions.

Most buyers compiled their sets by buying a few pieces at a time, and by 1964 only 20 per cent of Crown Lynn's New Zealand turnover was in complete boxed sets. By 1965 Crown Lynn had nearly 30 different replaceable designs. Promotional material that year – in exciting full colour – featured old favourites Autumn Splendour, Green Bamboo and Capri, as well as Shibui,

Autumn Splendour and Green Bamboo were the best-selling dinnerware patterns throughout much of the 1960s.

(Right) Topaz and Sapphire were added to the replacement range in 1965. This teapot in the Topaz pattern was also glazed in plain colours and decorated in other popular designs such as Autumn Splendour.

(Below) Sapphire oven-to-table ware from the 1960s or early 1970s.

New York, Narvik, Carousel and Fabrique. Buyers were urged to mix and match the plain colours of the Capri range, or add plain-coloured cups to the jazzy New York design. The next year, advertising was directed at the man of the house, suggesting that a 20-piece dinner set would make an ideal Christmas gift for his wife.

The replacement range was energetically promoted, with full-page advertisements in the *New Zealand Woman's Weekly* and other magazines. The 'internationally famous' chef Dennis Bone was enlisted to use Green Bamboo Ware as he showed his audience how to whip up exotic dishes like chicken chasseur and sauté of kidney turbigo.

Ultimately, the replacement policy became a burden. Designs had to be available for a minimum of five years, and could only be withdrawn after retailers had been given a year's notice. This made it difficult to withdraw old patterns, leaving less leeway to bring in new, modern designs. The department stores which stocked the Roydon and Kelston Ware brands weren't keen on the replacement policy. They preferred to sell dinner sets in fresh new designs.

THE IMMIGRANT EXPERTS / Through the 1950s and into the 1960s Tom Clark continued to hire experts from the other side of the world. The story of Ray and Eileen Machin from Stoke-on-Trent was typical of the 10 or 15 technical experts who came to New Zealand during this period. Ray Machin was a mould maker at J&G Meakin, and his wife Eileen was a hand decorator at nearby Johnson Bros when they spotted an advertisement in the local newspaper. An unknown pottery called Crown Lynn was looking for people to work in New Zealand. Almost all of Eileen's family – her mother, four sisters and a brother – lived nearby and worked at the Stoke-on-Trent potteries, so it was a big decision to emigrate to the other side of the world. But Crown Lynn had made a very attractive offer and they packed up and left for New Zealand in 1959. The air trip took five days; they remembered stops at Middle Eastern airports guarded by armed men. In those days travel was an important event and they dressed in their best clothes. Eileen wore her wedding hat.

The weary pair finally arrived at Whenuapai airport in Auckland at 11p.m. They were met by Tom Clark who – 'you know Tom' – immediately whisked them off to the factory for a midnight inspection tour. In the early hours they were finally dropped off at the house provided by the company. The next day was disappointing. Clark had arranged for them to be shown round Auckland, but no one turned up to drive them. However, it didn't work out so badly; they walked into town and joined in the celebrations for the opening of the Auckland Harbour Bridge.

Ray's first weeks at Crown Lynn were not easy. At that stage English techniques were streets ahead and he could see room for major improvements. Even simple things weren't right – cup handles were often crooked. The workers didn't take kindly to Ray's suggestions and the atmosphere was sometimes tense. For several months the couple seriously considered moving on to Australia where the Johnson Bros pottery

Mould making was a big job at Crown Lynn. This is the jug making department.

was keen to have them, but in time things improved, and Ray and Eileen settled into Crown Lynn life. After only two weeks Ray was promoted to mould room foreman. Tom Clark later described Ray Machin as 'a revolution in the place; he made things possible that previously had been absolutely impossible'. Eileen, an accomplished hand decorator, worked closely with head designer David Jenkin. Along with other hand decorators, she often demonstrated her craft at A&P shows and exhibitions.

A 1964 magazine article shows Ray and Eileen both happily at work in their respective departments, barely 20 metres apart. Ray enjoyed his new job and stayed with Crown

Lynn for the rest of his working life. Eileen changed jobs a few times, but she, too, spent many years in the ceramics industry.

After the first difficult months the Machins didn't regret the move. 'We never got rich out of Crown Lynn,' said Eileen. 'And we both worked, and worked hard. But we made friends. We had good times. We had the best time we've ever had in our lives … going to all the Crown Lynn balls and the soccer matches and God knows what, looking after all the couples that came out with their children – so we're auntie and uncle to 22 children.' For many years the English ex-pats held informal Sunday picnics with up to six families congregating at one of

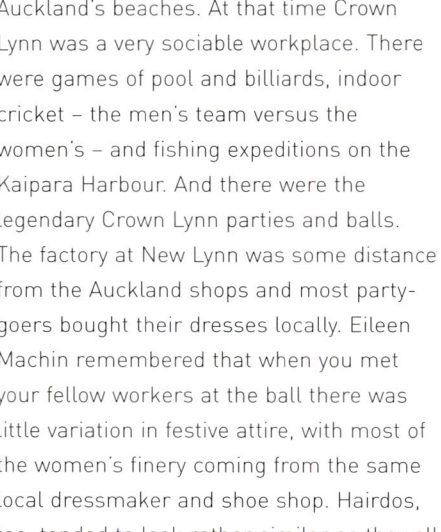

Vases hand painted by Eileen Machin in the 1960s.

Auckland's beaches. At that time Crown Lynn was a very sociable workplace. There were games of pool and billiards, indoor cricket – the men's team versus the women's – and fishing expeditions on the Kaipara Harbour. And there were the legendary Crown Lynn parties and balls. The factory at New Lynn was some distance from the Auckland shops and most party-goers bought their dresses locally. Eileen Machin remembered that when you met your fellow workers at the ball there was little variation in festive attire, with most of the women's finery coming from the same local dressmaker and shoe shop. Hairdos, too, tended to look rather similar as they all used the same – and only – local hairdresser.

For all the English immigrants it was a brave move to come to New Zealand. As long-time Crown Lynn employee John Heap pointed out. 'Half of them didn't even know whether we used knives and forks out here.' They were accustomed to wearing a collar and tie even in the engineering shop, and found the Kiwi jeans and T-shirt too informal for comfort. Some immigrants left Crown Lynn for new jobs, while others returned to England. Several immigrants including brothers Geoff and Gordon Ball stayed with Crown Lynn for decades. A few didn't fit in and either returned to England or found new jobs.

THE HAND-PAINTED VASES / Through the 1950s and 1960s Crown Lynn produced vases which were hand decorated with one-off patterns. Some of these were painted by newcomer Eileen Machin, some by other hand decorators, and some by Frank Carpay, before he left Crown Lynn in 1956. None of these designs were mass-produced, because Crown Lynn at that time did not have the technology to do so. Some vases were made by hand potter Daniel Steenstra; others were slipcast.

THE YOUNG CADETS / As a way to get new blood into his business, in 1961 Tom Clark introduced a cadetship scheme. Crown Lynn enlisted 'keen young guys' and sponsored them through ceramics degree courses at Stoke-on-Trent in England. They were bonded to the company for three years, but many stayed for longer. Chris Harvey began his career as a Crown Lynn cadet in 1967 and remained with the company until it closed nearly 20 years later, by which time he had done a stint as general manager. As well as raising the standard of management in Crown Lynn, these young graduates reduced Tom Clark's dependence on craftsmen from the English potteries. As he had done himself in his younger days, Clark made sure that the cadets spent time working on the factory floor, eyeball to eyeball with 'the real people who did things'. Studying in England was an eye-opener for the young men. Among the cadets were Tom Clark's son Geoffrey Clark, Ernie Cooper and John Homer, who reported that in Stoke-on-Trent everyone seemed to live and talk pottery. In 1964 Stoke alone had 320 potteries, though many were only one-

room enterprises. The smaller potteries were being swept aside as the big groups moved in and took them over. Only a few years later Crown Lynn made a similar move in New Zealand, gobbling up smaller potteries like Luke Adams and Titian Potteries.

HOLYOAKE SAVES THE DAY / In the months after Walter Nash introduced temporary import restrictions in 1959, Crown Lynn's fortunes improved greatly. However, the company was by no means secure; a return to unrestricted imports would cause a disastrous decline in the domestic market. At the time the pros and cons of import restrictions were much debated in the media.

Many New Zealanders and politicians were very keen to foster local industry to provide jobs. No one wanted a return to the days of the Depression, when a quarter of the workforce was unemployed. The crockery importers said that import restrictions would give Crown Lynn an unfair monopoly and force up the cost of household china, but Tom Clark said that he could not make sound business decisions in a 'busy this week, broke the next' environment. The turning point came in 1960 when an electioneering Keith Holyoake visited the factory and asked Clark what he needed to keep the business viable. Top of the wish list were higher tariffs on imported goods and tighter import

73

controls. Holyoake assured Clark that as soon as he won the election – and he was very confident he was going to win – he would set up a tariff board and Crown Lynn would be one of the first to be heard. Soon after the election Clark was invited to appear before the newly established board. He had been preparing for this moment for months and won his case against strong opposition from importers. Crown Lynn would continue to be protected by tariffs on imported china. Holyoake confidently predicted that the 1960s would be boom years for New Zealand manufacturing.

For Crown Lynn the imposition of tariffs was the reassurance needed to shift from art pottery and short runs – anything to make a few dollars – into full mass production. The factory had a solid technical base, good designers, a stable workforce and new machinery. It was well and truly ready to take advantage of the business opportunities that import controls offered.

This was not the last time Tom Clark had to go into battle over this issue. One of his most important jobs was to ensure that import controls and tariffs were retained. The restrictions applied only to the types of product that Crown Lynn was making – earthenware and vitrified hotel ware. Bone china could still be imported freely. There were always anxious times when there was a change of government, and periodically the government of the time reviewed the amount of crockery allowed into New Zealand. Each time the importers lobbied for more, and Crown Lynn lobbied for less. At one stage Crown Lynn employed a permanent lobbyist in Wellington to look after its interests.

Crown Lynn china was more expensive to produce than everyday ware from Johnson Bros and Meakin. New Zealand wages were higher and as a new business Crown Lynn had invested heavily in new machinery and in research and development. As the sole large china manufacturer in New Zealand, Crown Lynn also produced a much more varied range than its English counterparts. It was unheard of for an English factory to manufacture both earthenware and vitrified porcelain, but Crown Lynn made both. Fortunately for Crown Lynn, tariffs imposed on imported ware kept New Zealand-made china competitively priced but this was at the expense of the New Zealand public, who were forced to pay more for their tableware than they would have in an open market.

For Crown Lynn, as well as its customers, the protection of import controls was always a double-edged sword. Importers argued that Crown Lynn wasn't making a full range of products, and to satisfy the government watchdogs they were forced to make slow-selling uneconomic items like mustard pots, gravy boats, and salt and pepper sets.

Sometimes, too, retailers had a long wait for orders to be filled. After import restrictions were imposed, Crown Lynn could not meet the sudden increase in demand. In 1959 a retailer complained that he had to wait for up to two years for cups and saucers and plates, and a full year before a consignment of small basins was delivered. In 1963 the time lag between order and delivery averaged 10 weeks and could be as much as 16. Inefficient rail transport compounded the problem. It took three weeks for consignments from Auckland to

get to the South Island. And Crown Lynn was under constant pressure to develop new shapes and new designs. Tom Clark remembered that the buyers from major outlets such as McKenzies would always demand 'what's new, what's new'– and 'if you're trying to sell the same old thing, you've got no show'.

THE FACTORY TOURS / Once it was back on its feet, Crown Lynn looked for ways to raise its profile, partly to engage support for its campaign to retain import restrictions. In 1961 the company began running factory tours, which proved hugely popular. Personnel manager Trevor Lawrence hired part-time guides, and an estimated 150,000 people visited the factory in 25 years. Numbers peaked around 1966 with 20,000 visitors a year, but this proved too much of a strain on the guides so visits were cut back to more manageable numbers. In the heyday of factory tours, busloads of school children and women from Country Women's Institutes and farming clubs throughout New Zealand would arrive at Crown Lynn for a full conducted tour. Before the ladies rejoined the bus they were herded into the seconds shop where they would be encouraged to empty their purses with purchases of bargain-price crockery.

NEW TECHNOLOGY / By 1960 the Crown Lynn laboratory was in serious need of a ceramics expert. Denis McClure, the original chemist, had left and Clark needed an expert who understood the composition of the different clays and other raw materials used in the factory. Importantly, he had to know how these clays should be put together

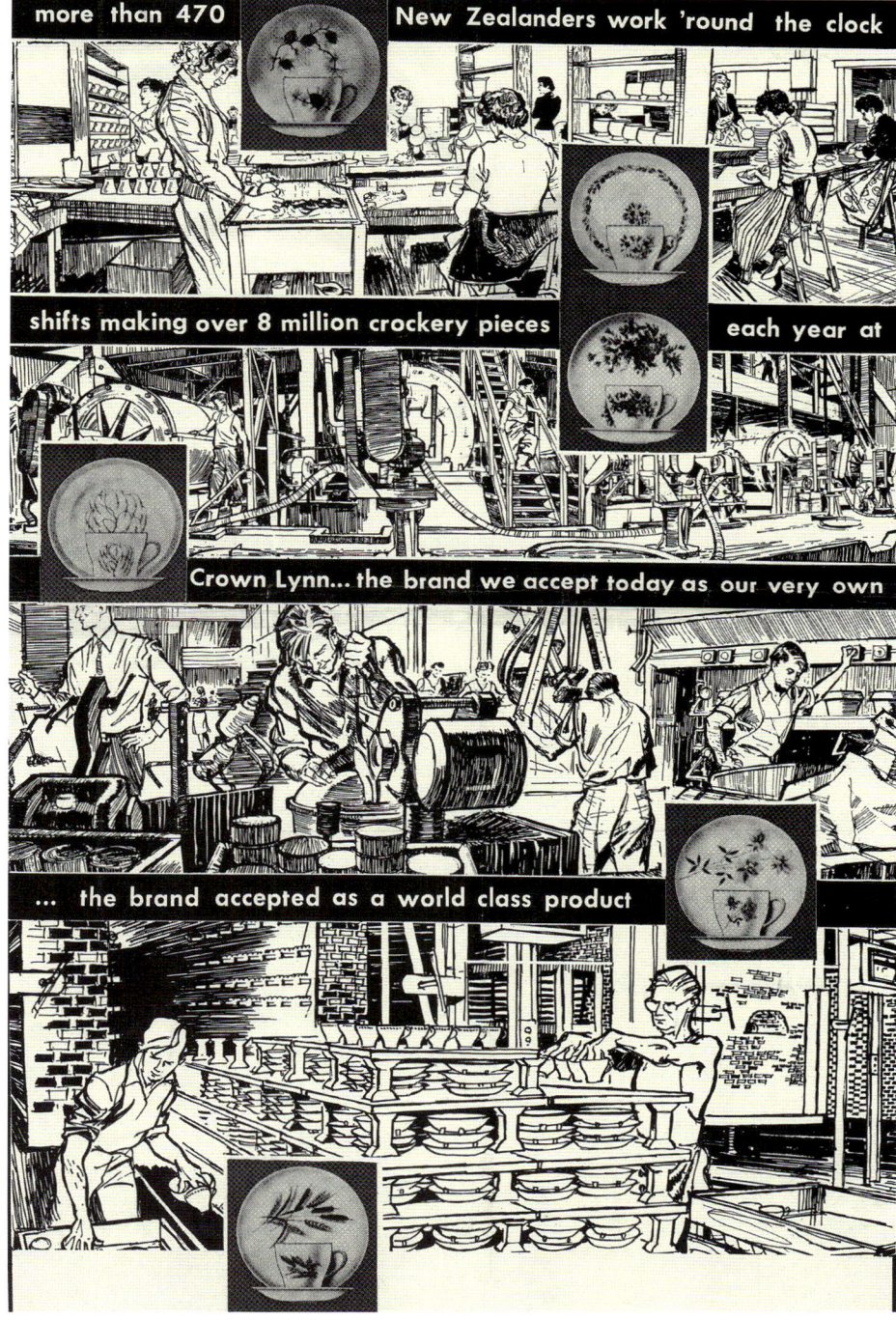

An advertisement from the early 1960s.

to make the best possible final product. An advertisement in the Stoke-on-Trent newspapers attracted Harry Jones, a ceramicist at Royal Doulton, and he brought his family to New Zealand in 1961 He was hired to improve and standardise mixes and processes, so that the end product was attractive, durable and consistent. That meant endless tests, mixing different proportions of different clays, firing them, then testing and analysing the results.

Glazes, too, needed attention. It was difficult to get the colours and textures right, and more difficult still to get consistent results every time. There are many Crown Lynn products with cryptic numbers and letters scratched or pencilled into their bases. These are the marks of the glaze makers, testing and retesting their formulas. As an added challenge many of the best and brightest glaze colours are made with heavy metals, such as lead, cadmium and copper. If the glaze is not prepared and fired correctly, it can lose its impervious surface, allowing toxic heavy metals to leach out. There was a story about an American family who kept their orange juice in a jug in the fridge. Over time, minute amounts of lead leached into the juice and poisoned the family. To avoid catastrophes of this sort, the Crown Lynn laboratory tested tableware for traces of lead and other chemicals. There was risk both from some glazes and some coloured lithographs.

THE COOK & SERVE RANGE / One of Harry Jones' first big jobs was to develop a new range of oven-to-table ware, the Cook & Serve range. Hundreds of clay mixes and glazes were tested in search of a product that would resist the heat of a domestic oven as well as sudden changes in temperature. The most promising samples were subjected to horrendous heat-shock tests. The scientists built a trapdoor into the bottom of a domestic oven so they could drop very hot items directly into cold water. By early 1963 Tom Clark was able to announce that Crown Lynn products performed as well as or better than their English, European and Japanese equivalents.

Clark's confidence was borne out by Mrs T. C. Walker from Kaitaia, who wrote to Crown Lynn after her Cook & Serve dish survived a fire in her stove: 'In the warming drawer directly above the oven I had one of your oven-proof cook and serve dishes with the leaf pattern on the base. When the fire brigade extinguished the fire, they carried the dish outside and put it on the wet grass. The dish did not break. Neither did it get one single crack or mark on it, as do most china dishes if they are over exposed to heat. I am still using this dish now and was thrilled to bits to find it had not marked. Consequently, all ovenware I purchase now will bear the Crown Lynn marking.' The new range included oval baking dishes, ramekins, lidded casseroles, dinner and lunch plates, coffee pots, cups, saucers and ashets (oval plates). There were also spice jars and kitchen containers, some with wooden lids. The first designs were Narvik, Vision, Blue Tango and Green Bamboo.

A NEW DINNERWARE SHAPE / In 1963, just after the introduction of the Cook & Serve range, a completely new shape in dinnerware was introduced, along with 20 new dinner set patterns, many designed by New Zealand artists. In the same year Crown Lynn brought out a new lightweight vitrified coffee cup and saucer in three patterns, and complete condiment sets – salts, peppers, mustards – in five colours and six patterns. From the artware department there were 12 new vases in matt white and black and a range of hand-turned TV lamp bases. Black and white television had arrived in New Zealand in the late 1950s, and it was customary to place a lamp on or near the appliance to cut down the effect of the glare.

THE COFFEE CANS / One of the popular innovations of 1963 was the new 'coffee can', the forerunner of the coffee mug. The coffee can had straight sides in contrast to the curved tea cup, but it still sat on a saucer. The new range included Allegro, Image, Blue Tango and Mogambo. Soon Tacoma, Bermuda and Saraband were added to the popular range. Mogambo was at first hand painted; later a transfer of the decoration was made. This design achieved instant fame when Gregg's Instant Coffee chose it for a series of advertisements.

TRADITIONAL DESIGNS / Throughout its life, Crown Lynn made extensive use of English-style decorative lithographs. Variations on the floral theme were always popular, as were 'romance' scenes featuring crinoline-gowned ladies and opulently dressed noblemen. These designs were used on cake plates, ornaments, sugar pots and cream jugs sold through McKenzies department stores. In 1964 Carl Ostmann from New York designers Commercial Decal told Crown Lynn that the most enduring design, featuring an elegant lady sitting on

GREGG'S GOOD COFFEE IS GOOD COMPANY

No wonder Gregg's is enjoyed in more New Zealand homes than ANY other instant coffee. It leads the swing because it sets the standard. There's nothing like the rich, hearty flavour that comes from the exclusive Gregg's master-blend of choicest coffees . . . nothing like the stimulating, refreshing satisfaction each friendly cup affords. Make Gregg's in a cup for yourself . . . in a coffee pot when you have company. It's the favoured form of hospitality today.

Gregg's INSTANT COFFEE NEW ZEALAND'S TOP SELLER!

58

(Right) **The Mogambo coffee can.**

(Above) **From the 1950s the popular 'romance' lithographs were used to decorate cake plates and other ware. This little jug was probably made in the 1970s. A few of the cake plates decorated with romance scenes or fruit lithographs were fitted with a central handle.**

(Opposite page) **The English-made decorations on these Crown Lynn plates were also used by British potteries, such as Old Foley, Royal Winton and Elijah Cotton (Nelson Ware).**

the grass, was first produced in 1904 and, worldwide, there had been 290 re-editions since then, each with a print run of 40,000 to 60,000.

Because most Crown Lynn products were decorated with lithographs imported from England or the United States, an imported English plate and the Crown Lynn equivalent sometimes carried the same decoration. The designs were generally unadventurous.

HOME-GROWN DESIGN / 'Solid colours, overall patterns and other contemporary designs are challenging the rose for her crown as queen of the New Zealand pottery market. A revolution is unlikely – the rose is too well entrenched for that – but more and more of her erstwhile supporters are deserting her.' *Femina* magazine, August 1959.

By the late 1950s Tom Clark was weary of imported lithographs with their endless variations on the romance scene and garlands of pink rosebuds. Crown Lynn was ready to break its dependence on traditional overseas designs and 'launch out boldly toward a distinctive New Zealand idiom'. He saw this as a means of increasing market share in New Zealand and also injecting new energy into the export trade. At the time most New Zealanders still identified closely with their English heritage, but Clark had noted a new trend – the more adventurous

New Zealanders were interested in interior design and were looking for new shapes and decorations. Crown Lynn's hand painters were already creating bold modern designs on vases and bowls, but these were one-off works and Clark was seeking similar innovation in his mass-produced tableware. To amass a pool of home-grown designs, Crown Lynn established a nationwide competition with a total prize purse of £610. Five hundred entries flooded in, from professional designers to enthusiastic amateurs. First prize was won by Wellington designer Otway Josling, whose simple crossed fernleaf pattern Reflections was put into production and sold well for years.

Domestic ware had a teal background, while hotel ware had mushroom. A 36-piece set was presented to Prime Minister Keith Holyoake, and the same pattern was later chosen by the airline TEAL for its in-flight tableware. The competition runner-up, Don Mills, submitted the Scandinavian-style design Narvik, which also proved popular.

At the time 'design' was not a well-understood concept in New Zealand. Crown Lynn was one of only a handful of industries that employed a designer, and there was little concept of a New Zealand style. Even illustrations of New Zealand flora and fauna were unusual in the early days, since all lithographs were bought from overseas. In

1961 two artists were commissioned to produce original designs with a New Zealand flavour. Art teacher Jack Crippen created eight new illustrations of New Zealand flowers and plants, while design-competition winner Bruce Bryant produced six native bird designs, including the tui, piwakawaka, and New Zealand parakeet. Crown Lynn did not have the technology to reproduce detailed full-colour illustrations, and the lithographs for these designs were made overseas.

In some cases competition also-ran designs proved more useful than the placegetters. Blue Tango, a 1961 entry by Emilie Beuth, did not win a prize but was used on domestic ovenware and a straight-sided coffee can.

1960s

Later, dinnerware in the Tango pattern was used in all 14 New Zealand Government tourist hotels – blue for the North Island and burgundy for the South Island. The first set went into the Chateau Tongariro in June 1966. Although Crown Lynn had copyright of her design, it is likely that Miss Beuth was compensated when it was adopted for such a large and profitable range. The design competition's entry conditions gave Crown Lynn the right to use every entry as they saw fit.

As well as giving Crown Lynn a supply of new designs, the annual competition raised the company's credibility and public profile. The presentation ceremony was a significant event in the Wellington social calendar,

reported on the television news and attended by powerful politicians and the design world elite. This gave Tom Clark and his colleagues a very real chance to impress the people that mattered with their products, their style – and their need for the retention of tariffs and import controls. On one occasion Clark departed from his prepared script to give politicians 'once round the bathroom with the razor strop' over their perceived failure to support local industry. Guest speakers, too, sometimes made their views known. In 1969 Prime Minister Keith Holyoake offered his own opinion on design. Whether a tractor or a dinner service, he said, a well-designed article should look good and do the job. 'A tractor should be a tractor and a plate a plate – not an imitation lettuce leaf.' New Zealand had definitely moved on from the Fancy Fayre Salad Ware of the 1950s.

The designs themselves were only part of the story when it came to developing a distinctive New Zealand style. To use the designs Crown Lynn needed new technology.

The installation of the first Murray Curvex machine around 1959–1960 made it possible to print new designs directly onto dinnerware. At first the engineers struggled to get the machine working, and it could print in only one colour, but later models could apply up to three colours. The designs were printed onto bisqueware, then coated with a protective layer of glaze and sent through the kiln again. This made the designs resistant to the acid detergents used in domestic dishwashers, which had begun to appear in the more affluent New Zealand homes. By 1963, in another important advance, Crown Lynn could make silk-screen

transfers using New Zealand designs. This was a much more laborious process than the Murray Curvex machine, but it too allowed designers to use home-grown designs rather than imported lithographs and make better use of colour.

TWO TOP DESIGNERS / Pivotal to the drive for better design was long-time head designer David Jenkin, who quietly oversaw the artistic direction of the pottery. He created many of the Murray Curvex patterns and silk-screen transfer decorations from scratch, and adapted competition entries to meet the requirements of mass production. Yet he was never one of Crown Lynn's 'characters'. Other designers, especially Frank Carpay with his Handwerk series, achieved prominence, but Jenkin's quiet influence was much more far-reaching and long-lasting. In 1968 another influential designer was invited to join Crown Lynn. Mark Cleverley, who had already won two Crown Lynn design competitions and had been well placed in five more, was hired as 'development designer'. He was to examine world trends and then produce designs in keeping with overseas tastes, but with a distinctive New Zealand influence. Cleverley recalls that his work was very satisfying. 'We'd have a lot of laughs, a lot of fun. It was a big game.'

THE ASSEMBLY LINE / Crown Lynn relied heavily on production-line labour. Workers sat at benches all day, every day, painting lines on saucers or fixing handles to cups. To alleviate the boredom some sang as they worked. In the 1960s tea cups were Crown

Lynn's best-selling item, with over one million churned out in a single year. In 1969 as many as 14,000 cups and saucers could be produced in a day. A large percentage were sold in plain white to company cafeterias, sports clubs, community halls and marae.

Despite steady progress towards mechanisation, much of the production work was still done by hand. A cup took six days to make and from raw clay to finished item it would pass through up to 15 pairs of hands. Each cup was made individually in a moulding machine, put in a drier for a set period, then it was fettled – the smoothing down of rough edges – before the handle was attached. A quick worker could attach 4300 handles in an eight-hour working day. After another firing each cup was backstamped and decorated, then loaded into a kiln for a final firing. They were checked for quality and any with defects went to the seconds shop and the rest were sent to the warehouse for packing and distribution. The workers – mainly women – in these repetitive jobs were not well paid, but they could top up their earnings with performance bonuses and overtime. Some stayed for decades; Mrs Maude Leadbeater and Mrs Barbara Mulcaho were photographed in 1966 after nearly 20 years' service. The fettling and cup-handle

departments were overseen for many years by Ringi Ngakuru, one of Crown Lynn's valued staff. She ran a tight ship, recalled Chris Harvey: 'One day the personnel manager arrived with a new girl, a lovely young lady, who was made up and had big long fingernails. The girls had to put their hands in the cups to flip them over, and Ringi said you can't work here like that. She had a pair of big scissors and grabbed her hand and went chop chop chop – and this girl was ready to go to work.'

THE DECORATORS / Much of Crown Lynn's decorating was done by hand. A 1965 issue of the in-house magazine featured newly appointed manager Maude Bowles, who was responsible for a department that decorated 70,000 items each week. With a staff of 30 and a string of new workers to train, this was a substantial job. The decorators applied lithographs, silk-screen transfers, coloured bands and gold lines to the ware. They were also responsible for hand painting, which at that time often involved endless repetitions

83

This ornate urn, made especially for the Queen, was not presented to her. Glazed in plain white, the urn was also sold as a souvenir of the Royal visit.

of the popular Fleurette design that was sold through the Milne and Choyce, and DIC department stores. Fleurette was a brazen copy of the popular English Belle Fiore pattern. One of the chain stores asked Crown Lynn to copy the pattern and for a while the words Brereton Ware were added to the backstamp in recognition of DIC chief buyer Jack Brereton. Painting Fleurette was a production-line job, with six painters each applying a single colour. Fleurette was introduced in the late 1950s and it was still being made in 1979.

THE QUEEN'S VISIT / 'All the snooty ladies who wouldn t have anything to do with Crown Lynn now bought it. We'd arrived.' Tom Clark.

A visit from the young Queen Elizabeth II in February 1963 was a major turning point for Crown Lynn. Thanks in part to import controls Crown Lynn's products were selling well, but most middle- and upper-income New Zealanders still aspired to English china. The Royal visit changed attitudes almost overnight. In Tom Clark's words the visit 'deodorised the whole stink of the past'.

The Queen – described in the local media as 'fresh and cool as a summer breeze' – kept to a strenuous schedule. She toured the Tattersfield carpet factory in the morning and Crown Lynn in the afternoon. Tom Clark escorted her through the entire factory and introduced her to the departmental managers including bisque warehouse manager Joyce Maunder and David Jenkin, who had changed from his customary shorts and open-neck shirt into more formal clothing with just minutes to spare. Throughout the tour a retinue of photographers and journalists from Australia, the United States and England jostled for position, but despite the melee not a single item was broken.

The visit was rather rushed. Eileen Machin recalled that hand-painted ware was put on display for the Royal visitor, but she had time for only a quick glance. Worse, the Queen showed no interest in an ornately decorated urn which had been specially made to present to her. However, she did take the time to tell factory manager Fred Hoffman that she found the development of new machinery one of the most interesting aspects of the pottery. When the Queen passed through the cup-handle department, workers couldn't resist glancing up, but continued automatically to cut, trim and fasten handles to cups.

Outside the main entrance the Queen met all eight members of Crown Lynn's Board of Directors in front of a grandstand specially erected to seat 480 friends and relatives of Crown Lynn employees. The factory emptied in seconds when the staff got the signal that they could farewell their visitor from out the front.

That evening Tom Clark was invited to dinner on board the Royal yacht. Afterwards he recalled that the young Queen was delightful company, interested in everything and very much in love with her husband, the Duke of Edinburgh. 'She was marvellous . . . she's looking up at the Duke. And her feet are killing her. She's been out all day and her feet are killing her. And she slipped out of her shoes . . . and she's just looking up at the Duke . . . Oh yes . . . he was definitely the cat's bloody pyjamas.'

THE EXPORT DRIVE / In the early 1960s New Zealand was riding a wave of full employment and optimism. With a stable domestic market Crown Lynn needed to export if it was to continue to expand, and its managers criss-crossed the world in search of new markets. Australia was the first target. In a symbolic ceremony in 1961, overseas trade minister Jack Marshall was invited to stamp the magic word 'Export' on the first case of dinnerware destined for an overseas market. The 'bold venture into the unknown' soon paid off. By 1964 Crown Lynn had 130 Australian outlets. Australian sales manager Jack Mason recalled that getting into the market was difficult: 'We had to

battle hard in those early years, particularly with the Aussie buyers who thought that dinnerware could only be made by British potteries. However, with quality, advertising and the replacement policy we finally wore them down.' The dinner sets Reflections, Narvik, Fabrique and the plain-coloured Capri were top-selling patterns in Australia. Most were branded as Kelston Ware rather than Crown Lynn. In a major coup in 1968 Crown Lynn won a contract to supply the Australian armed forces with monogrammed tableware – a total of 121,620 items. An Australian dealer came up with an interesting reason for Crown Lynn's success: 'Crown Lynn is easy to pronounce, more than can be said for some of the European brands.' During the early 1960s outlets were also established throughout the Pacific Islands, in Fiji, Samoa and American Samoa, Tahiti, Noumea, New Guinea and the Cook Islands. Setting up shop in 'steamy Tahiti' was a struggle for company representative Denis Beggs, who said he would have much preferred to be there on holiday. Australia and some Pacific Islands had their own forms of import protection, forcing Crown Lynn to pay tariffs.

Crown Lynn was also making inroads into the North American market. In 1964 the company announced that for the first time ever New Zealand china would soon be sold in North America. Competition was tough – one store stocked 700 different dinnerware patterns from all over the world. Everyone knew Meakin, Johnson Bros and Doulton, but potential distributors wouldn't believe that a little country like New Zealand could meet high-volume orders for good-quality

china. And Crown Lynn didn't always get things right. Alan Topham remembered one trade fair in snow-covered Atlanta where crowds of retailers flocked to an English stand showing brightly coloured Portuguese designs – while Crown Lynn was all but ignored.

The first ware to be sold in Canada was the Dorothy Thorpe range from 1966. This was quickly followed by middle-of-the-road patterns such as Tres Bon, Egmont and Hacienda. By the early 1970s Crown Lynn was among the leading imported brands selling in Canada, and in the United States Crown Lynn was being distributed by large department stores and the mail-order firms Sears Roebuck and J. C. Penney.

In the 12 months between 1966 and 1967 Crown Lynn's exports doubled. The biggest markets were Australia and Canada. More than one million pieces of tableware were shipped abroad, and Crown Lynn was among New Zealand's top export earners in the manufacturing sector. Trial consignments went to Japan, Hong Kong, Singapore and South Africa. By the end of the decade there were over 500 outlets in Australia and 150 in Canada. Crown Lynn was being sold in the Pacific Islands and South Africa, and had a small presence in Malaysia. Next on the list was Europe, though that market never really gained momentum apart from some sales in Greece.

It is true that Crown Lynn displayed considerable energy and innovation in its search for export markets, but the company was by no means operating in isolation. During the 1960s the government strongly encouraged New Zealand manufacturers

(Opposite page) Dorothy Thorpe's Santa Barbara. The occasional ball-handled coffee set turns up in New Zealand, but they are almost all defective in some way. The only ball-handled sets sold in this country were from seconds shops.

to export, and provided practical support through tax incentives. Even the Jaycees service club lent a hand, organising a major trade mission to Fiji in 1964. In 1969 the company received a special government export award and that year Tom Clark led a trade mission to South-East Asia to 'repay the government who stuck to us against all sorts of ill-informed criticism in the early days'. Clark believed that his company had to be big if it was to have any clout in the world marketplace. The larger the factory, the greater the capacity to meet large export orders – but this meant that more export orders had to be secured to absorb continuing production. With a huge amount of money invested in machinery, staff and materials, the factory had to be kept going at full production in order to be profitable. The kilns could not be easily shut down and had to be fed with new ware, day after day, week after week, month after month. And the company had to pay the staff who operated the machinery and fed the kilns – week after week, month after month, year after year.

DOROTHY THORPE / A pivotal aspect of Crown Lynn's export drive was its alliance with Californian glassware designer Dorothy Thorpe, who created a special range for sale in the United States and Canada. Thorpe, who was keen to get into the ceramics

market, listed among her clients a host of movie stars as well as Princess Grace of Monaco, the Shah of Iran, and Conrad Hilton's luxury hotel chain. Her Crown Lynn range was launched in North America in 1965, and the next year a modified version hit the New Zealand and Australian markets.

Tom Clark recalled that in the early days of the alliance Dorothy Thorpe spent a week or so at the pottery – 'very much like a gracious American lady visiting the natives'. The visit was of strong interest to the media, and Crown Lynn staff certainly recognised her importance. Marketing manager Alan Topham recalled that not long after Thorpe's arrival, she was standing talking to him outside the factory entrance. Cup-making supervisor Ringi Ngakuru came out of the building and down the steps, and presented her with a pounamu (greenstone) pendant. 'It was Ringi's way of saying "welcome; I hope this alliance will be good for us all".'

Thorpe created some very stylish shapes, with clean, sweeping lines and smooth curves. A new addition was the wide bowl 'for the tossed salads the Americans serve, almost as a matter or course, with their dinner'. The range for export had spectacular round 'ball' handles, which definitely set them apart from other products. However, at the time the Crown Lynn staff were appalled at the prospect of having to make

such difficult shapes. Mould maker Ray Machin remembered her designs as being nothing but trouble. 'She caused the biggest headache. She wanted a ball as a handle. When you fired them, they exploded in the kiln.' Machin told his boss that losses would be too heavy to make the range viable, but was told, 'You've got to make them, you've got to make them.'

They soon worked out that a small hole in the bottom of the ball handle would prevent explosions in the kiln, but the range was never particularly practical. The ball handles were difficult to grip and inclined to fall off at the slightest bump.

Soon after the Dorothy Thorpe range was released in the United States, it was adapted for the New Zealand and Australian markets. The ball handles were done away with, but the cups remained wide and boat-shaped. They were certainly elegant but the general opinion was that they were pretty much useless – your tea went cold too quickly. Then there were the striking but top-heavy coffee pots that would tip over at the wobble of a tray. They weren't a problem, said Alan Topham, 'because you never used the damn things anyway'.

Under Thorpe's rather pedantic influence, Crown Lynn's quality hit a new high. The chemists produced glazes that exactly matched her colours and the mould makers

Originally part of the ball-handled range, the Palm Springs design was made in more conservative shapes for the New Zealand market.

88

copied her shapes perfectly. She was impressed with the size of the Crown Lynn factory, the degree of mechanisation and the quality of the final product. Even the hit-or-miss backstamp process was improved. Previously, production staff simply whacked a stamp on the base of the ware without worrying too much about placement or legibility. Thorpe insisted that all backstamps were standardised and centred. This created another headache for the production team, only resolved by making backstamp transfers which were carefully applied by the decorators.

In the end, though, the Dorothy Thorpe experiment was not a huge success. In retrospect Tom Clark felt that he and his managers had been dazzled by Thorpe and her glamorous American connections – but she promised more than she delivered. Thorpe was responsible for selling the range in the United States and it turned out that she didn't have a solid marketing network. This failing, combined with currency changes, resulted in disappointing sales and the alliance slowly crumbled.

Today the Dorothy Thorpe range stands out as an elegantly designed and well-executed product. The Pine pattern, in particular, has proven enduringly popular, though its surface is vulnerable to knife marks and many a set has been relegated to the family bach because of the unpleasant jarring noise made by cutlery scraping over textured glaze. The Dorothy Thorpe range was a collaboration between New Zealand and American design. The shapes were all designed by Thorpe, along with four decorations – Brocade, Monterey, Santa Barbara and Laguna. The two remaining designs, Palm Springs and Pine, were both created by New Zealanders. Palm Springs was a prize-winning design by Mark Cleverley and his name appears on the backstamp. Originally designed in black on dark grey, the colours were changed for mass production. Ware in the Pine design is also marked as being designed by Mark Cleverley, but this is not accurate. The design was created by David Jenkin, but was attributed to Cleverley by mistake.

The demise of the Dorothy Thorpe range was not the end for the Pine and Palm

Springs designs. Dinner sets sold well in New Zealand and Australia, and as late as 1968 cups and saucers in these patterns appeared in the Shape twenty-5 range. Aspects of Thorpe's designs also endured, particularly in the shape of the serving bowls for Egmont and Yucatan dinnerware.

THE GRANULE TECHNIQUE / One of the technical breakthroughs of the Dorothy Thorpe era was used to great effect for many years. Referred to as the 'granule technique', it created a textured effect on the finished ware. A Malkin automatic printing machine stamped a pattern in glue on an already glazed and fired plate. Fine granules of glaze were then sprinkled across the plate, sticking to the glued areas. When the plate was fired again, the granules melted into the glaze to form a raised pattern. Developed by the designers with the aid of chemists Harry Jones and Reg Taylor, the technique was used on the Pine pattern, and later in other designs such as David Jenkin's Egmont.

AIR NEW ZEALAND / In 1965 Crown Lynn supplied tableware to the government-owned Air New Zealand for its new DC8 jet airliners. Designing tableware for the national airline was a very high-profile job and Crown Lynn rose to the challenge with a smart Maori design in brown on a turquoise background. The cups had to be specially designed to stack neatly and the whole range was in durable but lightweight vitrified ware. At the time Crown Lynn tried hard to persuade the airline to use the design in gold on turquoise and a few samples in this colour still exist. However, in the end they decided on brown

(Top) **Pine.**

(Bottom) **A butter dish from the Air New Zealand range.**

Roydon Tiny Tots.

90 because it is much more durable than gold. Throughout the next decade Crown Lynn also made ashtrays for TEAL airlines and beakers for Air New Zealand in golden brown and green.

DESIGN HONOURS / By the mid-1960s Crown Lynn was beginning to get noticed in the design world. In 1966 Auckland florist Berin Spiro took three Crown Lynn vases to London to demonstrate his craft before an audience of 1000 florists. The vases, crafted by Tam Mitchell, were decorated with a Maori pattern in relief and glazed in a rich reddish-brown colour. The largest was close to a metre high. It is most likely that they were left in London after the exhibition. The Yucatan dinner set, designed by Robert Drake, was displayed at the 1968 Rothmans Industrial Design Awards, and Crown Lynn tableware was included in a 1969 display at the Design Centre in London.

NURSERY WARE / Crown Lynn had made its first attempts to break into the nursery ware market in the early 1950s. At that time shop shelves were dominated by the expensive Royal Doulton Bunnykins series. Crown Lynn's selection, decorated with imported transfers, was more affordable but seen as second best. Among the first products were plates, and cups and saucers decorated with nursery rhyme and teddy bear themes. By the 1960s Crown Lynn's nursery products were well established. In December 1964 the company was promoting the Bunny and Wee Pets ranges as ideal Christmas gifts. The Roydon Tiny Tots range made in the mid-1960s for McKenzies chain stores featured various imported cat transfers. These proved as popular with adult cat enthusiasts as they were with children.

In 1966 Shirley and Sam Lawson received nursery ware gifts for their new quintuplets and older sister Leeann. The Lawsons met when they were both working at Crown Lynn. The factory continued to make various patterns of nursery ware through to the 1980s.

SELLING CROWN LYNN IN NEW ZEALAND / Throughout the 1960s there seemed no limit to how big Crown Lynn could grow. The New Zealand economy was surging ahead and, in an optimistic and adventurous frame of mind, householders were eagerly buying homeware to suit their newly built or freshly redecorated homes. There was increasing interest in a more modern style, which had first emerged in Swinging London with its mini-skirts, Mod fashions and beehive hairdos. Crown Lynn's sales, both export and domestic, soared. Tom Clark's aim was to sell his product in every New Zealand town – big or small. Once a year the growing band of Crown Lynn retailers was brought together for a pep talk and a look at new products. The successful traders got awards, the less successful a firm instruction to do better. During the 1960s Crown Lynn held annual window-dressing competitions with substantial prizes encouraging retailers to ever more spectacular flights of fancy. In 1964 the £60 first prize went to Taumarunui Hardware, whose display featured a Mini Minor car with each wheel balanced on a single upturned cup. 'The handle is as strong as the cup', boasted a hand-lettered placard. The car had been jacked up and carefully lowered onto the cups. With Crown Lynn's reputation for poorly attached handles, this

display won favour with the judges.

The next year the shop put a live bull in their window, but managed only a third placing despite the animal's effectiveness as a crowd-puller. The bull, said the shop owner, had not been given tranquillisers. 'It has been entered in local school and A&P shows since birth and has led a most sheltered life.'

During this decade Crown Lynn never missed an opportunity for publicity. In 1963 Tom Clark presented Prime Minister Keith Holyoake with a specially made 204-piece dinner service for the newly completed New Zealand House in London. The next year Clark invited the country's top athletes to tour Crown Lynn, and local newspapers published photographs of Olympic runners Murray Halberg and Peter Snell – accompanied by their glamorous

wives – watching hand painters hard at work, and smiling at the jokes of factory manager Fred Hoffman. The wives were presented with Crown Lynn dinner sets. When beauty queen Helen Iggo competed in the Miss International contest in California, she used a Crown Lynn ceramic gourd to pour water from Lake Taupo into a 'pool of international friendship'. The televised ceremony was viewed by thousands of Americans, and New Zealand newspapers covered the event avidly.

By 1967 Crown Lynn was producing 10 million pieces of tableware and pottery every year, and more than half was sold in New Zealand. The best-selling designs included Autumn Splendour, Shibui, Capri, Golden Fall, Green Bamboo, New York, Narvik and Fabrique. A range of floral designs on lightly patterned backgrounds also sold well. Alan Topham, who had joined Crown Lynn in 1963,

Middle-of-the-road floral designs were consistently good sellers in New Zealand and overseas. From top left: Roydon Woodland, Devon Rose and Regal Rose. Below: Roydon Harvest. Exclusive to McKenzie chain stores, the Roydon brand name was derived from the McKenzie brothers' names – Roy and Don.

CROWN LYNN

(Above) **White Kelston Ware with a gold trim was a popular choice through the 1960s.**

(Opposite page) **Colourglaze Ware.**

was responsible for selling this enormous – and growing – output. At that time department stores were a major force in retailing. McKenzies had 53 stores and Woolworths 60, and Farmers Co-op also had a chain of stores throughout the country. Tom Clark decided that these substantial outlets should not sell ware with the Crown Lynn brand. Instead, they were given the Kelston Ware and Roydon brands. Clark argued that the Crown Lynn brand should be kept for the franchise shops and more upmarket department stores such as Milne and Choyce and DIC. This allowed him to guarantee that franchise shops were sole distributors in their own 'patch', yet he could still sell his

product in department stores under another brand. The department stores were allowed to choose their own exclusive lithograph designs. There was a huge demand for pretty floral dessert sets of six plates and a matching serving bowl. All three chains sold Kelston Ware in pure white, white with a gold band, and the plain-colour Colourglaze.

Crown Lynn's plain-coloured ware, sold under several brand names including Capri, Colourglaze and South Pacific, sold consistently well. In 1961 buyers were advised to 'Buy it in pieces, as a one colour set or gaily mix it together. Replacements are always available so you can add to and build on Capri as you wish.'

Popular dinnerware designs included Tam-O-Shanter, Aztec and Carousel, which was promoted as 'a whirl of gay bronze bands … a thrilling break from the conventional.' To keep retailers happy and on their toes, Crown Lynn sales reps Joe Elliot and Henry Sawyer travelled the length and breadth of New Zealand. The men were well known on their respective beats and Crown Lynn received hundreds of messages after Joe Elliot died suddenly in 1968. To raise public interest the company set demonstrations of hand painting or hand potting in shop windows. Decorators Doris Bird and Eileen Machin and hand potter Daniel Steenstra were popular demonstrators for many years. They gave away their work to anyone who was interested, though of course it wasn't fired so it didn't last.

One of Crown Lynn's biggest customers was the Government Stores Board, which bought millions of pieces of vitrified porcelain each year. The Stores Board controlled the

94

supplies for government departments, schools, hospitals, the armed forces, the Railways Corporation and the government-owned Tourist Hotel Corporation. In those days government departments supplied their substantial workforces with tea and biscuits in the staff tearoom, and it was in Crown Lynn's interests to ensure that every cuppa was sipped from a Crown Lynn cup and every biscuit served on a Crown Lynn plate. These were big orders. The Railways bought nearly 300,000 cups and saucers a year, and all New Zealand hospitals bought Crown Lynn vitrified ware in Colourglaze for staff use and various patterns for patients. Topham recalled that the Stores Board buyers Fred Lee and Colin Pringle were supportive of local business but required value for money.

The hospitality industry was another big market for Crown Lynn. Specialist distributors, most prominently Gibsons and Paterson (Gibpat) – which was bought by Crown Lynn in 1968 – sold vitrified ware to hotels, coffee bars, restaurants, marae and community organisations. Many bought the standard Colourglaze or plain whiteware, while others bought monogrammed ware. Crown Lynn dinnerware went into the new Kerridge building in Auckland's Queen Street, the White Heron Lodge in Parnell and a new luxury resort on Pakatoa Island in the Hauraki Gulf, serviced by the new hydrofoil

ferry. In 1968 the brand-new 14-storey Intercontinental Hotel in Auckland ordered specially designed dinnerware for both its restaurants. In the early days when Crown Lynn was still struggling to prove itself, George Tunnicliffe from Gibsons and Paterson developed the legendary 'bash test'. If a customer was wavering between English ware and Crown Lynn, he would pick up an English plate and bash it against a Crown Lynn plate until the weakest broke. Almost invariably Crown Lynn won and another sale would be stitched up.

Popular with low-income people, but also patronised by the more affluent, Crown Lynn seconds shops were also a substantial money-spinner. With production hitting 10 million pieces a year there were plenty of defective items coming off the production line. The worst were smashed – one worker remembers sending dozens of big white swans crashing into a rubbish pit in the early 1960s. However, it made good economic sense to sell off slightly imperfect items at a discount rather than dump them. By the mid-1960s Crown Lynn had four seconds shops in the Auckland area. The most active was right next to the factory, but shops in the then low-income suburbs of Ponsonby and Mangere also did well. As Crown Lynn's production grew, there was a risk that the New Zealand market would be flooded with seconds, and container loads were exported to the Pacific Islands. For many years the shops were run by Bill Wiseley, who set up the first temporary seconds shop in front of the factory in the early 1950s. Mary Stewart, who worked in the office in 1950, remembered the staff bringing out mistakes and experimental oddities to add to the ware laid out on trestle tables outside the factory. Like many jobs at Crown Lynn, grading for seconds was a tedious task, but some staff stuck at it for years. In 1968 Tom Clark presented Miss Anne Ward with a watch and bracelet on her retirement after 20 years as a grader.

THE SHAPE TWENTY-5 RANGE / The angular Shape twenty-5 range was introduced in 1968. Developed by designer David Jenkin and modeller Tam Mitchell, the new range featured a cup with straight sides and a tapered base – 'a style now particularly popular for both coffee and tea'. The new angular shapes were popular with the production staff because the flat surfaces could be decorated with silk-screen transfers. The previous curved cups could only be glazed in a single colour or decorated with lithographs. The range included coffee sets, cups and saucers, a gravy boat and stand, oven-proof casseroles and baking dishes, salt and pepper shakers, jugs – and a new

'coffee beaker' which was simply a coffee mug with a matching saucer. Among the first products in the new shape were coffee sets in the award-winning designs Novelle, Carnaby and Time Out. At the time Crown Lynn's forward orders weren't too promising and word went out that the new coffee sets should be promoted heavily. The result was

(Above) Hospital and hotel ware.

(Opposite page) From top: Carousel, Tam-O-Shanter and Aztec.

much better than expected; the sales team sold thousands throughout the country.

In the same year as Shape twenty-5 was launched, Crown Lynn brought out a new range of vitrified ware for hospitals, restaurants and hotels. Designed with export markets in mind, the range had 30 different items including six sizes of the standard milk jug. Still in demand was the 'invalid's cup' for disabled hospital patients.

MORE EXPANSION / In the late 1960s Crown Lynn was again looking to increase output. Overtime shifts were the norm and there was a need to use more automatic machines to keep labour costs down. Factory extensions were planned and factory manager Fred Hoffman and head ceramicist Harry Jones were sent overseas to investigate options. Their first stop was Japan, where they tested a new type of kiln by making, firing, glazing and then firing again hundreds of plates made from Crown Lynn clay which had been sent over specially for this purpose. Both men, said Jones, lost weight, working hard on an unfamiliar diet. 'Fred and I did all the loading and unloading of the kiln – six days a week. I was game to eat most of the Japanese food – no western-style grub in Miunami – but Fred did not help himself by subsisting mostly on Sapporo beer and sake.'

From Japan they travelled through the United States, England and on to Europe to look at German-manufactured plate-making machines in action – and to take in the Munich beer festival where Hoffman rolled up his trousers and jumped up onto the stage to join the band. Finally back home, the two globetrotters gave their recommendations for the new factory. The company bought new kilns, and machines for preparing clay, making cups and plates, and decorating. John Heap, who was assistant to Fred Hoffman for much of this time, recalled that a lot of this machinery had to be adapted to Crown Lynn requirements, and two difficult years passed before it was finally in full production. During the factory revamp a new electrical switchboard was installed, replacing the original which had been built for Ambrico in 1942 by electrician Ron Keene. In 1967 Keene said that Crown Lynn was one of the three biggest electricity consumers in the Waitemata, along with the Chelsea sugar factory and the Devonport naval base.

At the same time as Crown Lynn was expanding, its sister company, the porcelain division, was generating a steady turnover, making generally unspectacular electrical items. In 1964 it was making non-slip tiles to

96

From left: Novelle, Carnaby and Time Out.

edge steps, radio aerial insulators for taxis and cars, ceramic door knobs and finger panels, soap dishes and towel-rail ends, and even ceramic decorations for women's shoes. There was a steady market for the little 'H' and 'C' knobs for the tops of hot and cold domestic taps. Other oddities included shapes used for manufacturing rubber balloons and babies dummies. There were also big orders for the white reflective markers which were put into main roads up and down the country through the 1960s and 1970s. In 1967 the porcelain department introduced its new Feminine Approach range of door handles, keyhole plates and light-switch plates decorated in the Aztec design, and delicate floral and fleur-de-lis patterns. The range was later extended to include toilet-roll holders, soap dishes, cupboard handles, towel-rail holders and toothbrush stands, all made of dry-pressed porcelain.

When the range was first released, retailers throughout the country ordered almost 6000 pieces. Crown Lynn produced 1200 pieces a week and intended to export half that amount. However, the household items were only sidelines. Ninety per cent of the factory's output was for the electrical appliance industry. Bars for the elements of electric heaters, a pioneering product in the 1940s, were still in demand. Insulators and power-pole fuses for the electricity industry were still being made in 1979.

THE LYNNDALE RANGE / In 1967 the Lynndale range was introduced. Two patterns – Rose Red and Sierra Pine – were imported lithographs, while three – Hacienda, Staccato

(Above left) One of Crown Lynn's more novel products in the 1960s was the Down Town series, produced for Trillo's restaurant and convention centre in Auckland. Featuring a map of the city with hotels – and of course Trillo's – prominently marked, the design was used on all their dinnerware. The convention centre seated up to 1500 people, so this was quite a big order for Crown Lynn. Similar plates were produced with maps of Wellington, Dunedin and Christchurch.

(Above right) Soap dishes of this type were inset into the bathroom wall.

Titian was never a large enterprise, but its more artistic approach was an alternative to Tom Clark's focus on mass production. Titian began as a backyard operation but moved to commercial premises in Henderson in 1958. At that time Cameron and Dorothy's son Neil left Crown Lynn to join his family business. In the years before the takeover, Crown Lynn and Titian produced many items that looked very similar. Sometimes this was coincidence – both potteries copied an English product that was fashionable at the time. However, there are stories of Crown Lynn staff helping out at Titian without Tom Clark's knowledge. Crown Lynn mould maker Ray Machin remembered visiting Titian after dark to help out, especially in the early 1960s when he had just arrived from England and wasn't particularly happy at Crown Lynn. Legend has it that Crown Lynn's head designer David Jenkin also did some after-hours work for Titian, of course without telling Tom Clark. There is a story that Jenkin designed Titian's Swoose – an elegant wall vase which looks like a cross between a swan and a goose. One day Jenkin was called into Tom Clark's office and to his horror noticed a Swoose on the desk. He expected to be sacked on the spot. Instead, Clark said, 'Why can't we do stuff like this? Why can't we do what Cam Brown is doing?'

To fund a new factory, Titian Studio, now named Titian Potteries, became a publicly listed company in 1965 and began to produce once-fired earthy-coloured brown coffee ware, which was selling well at this time. Crown Lynn began to buy shares in this potential competitor and by 1967 was the major shareholder. In 1968 Crown

and Capistrano – were designed by David Jenkin. Sierra Pine was already being sold by Crown Lynn as Pinewood. The Lynndale range was promoted as youthful, colourful, exciting – and disposable. An advertisement told buyers that 'when you get tired of it, go on outside and smash the lot – then pick another set and keep it just as long'. The advertising campaign featured a full-colour photo of young people eating pasta. At the time most New Zealanders were still eating meat and three veg, and pasta was seen as a modern, exciting dish.

TITIAN STUDIO / Crown Lynn's only real rival in Auckland was Titian Studio, owned by Cameron and Dorothy Brown and their family. With a staff hovering around 20,

Lynn announced a 'close association' with Titian – in effect a takeover. From this date Crown Lynn's matt white art pottery was produced at Titian, freeing capacity at the New Lynn factory to mass-produce dinnerware. Through the next decade Titian was managed by Fred Hoffman, mainly producing whiteware and once-fired coffee sets, jugs and kitchenware in browns, earthy greens and a light brownish honey colour. Items were designed and moulds made at the Crown Lynn factory, and raw materials also came from Crown Lynn. There was some ill-feeling about the takeover and before long Cameron Brown left to set up a new pottery, Orzel Industries. Titian was sold off in the mid-1980s and before long the new owners closed it down.

THE WILLOW PATTERN / The 1960s was a time of innovation, but Crown Lynn never lost sight of the market for traditional designs and so introduced the Willow Pattern in 1968. In England this design was reproduced in dark cobalt blue, which is inclined to 'migrate' in the kiln – tiny specks float through the air and settle where they are not meant to. Crown Lynn got round this problem by using a paler blue. Later, the pattern was also produced in dark blue, and in black and white. The Willow Pattern is an adaptation of an eighteenth century Chinese design, illustrating a story of forbidden love. Beautiful Koong-se falls in love with Chang, a lowly clerk, and the pair run away together. The girl's father eventually tracks down the lovers and kills them, and they change into doves, symbolising constancy and immortality.

(Left) **In 1967 New Zealand's currency changed from pounds, shillings and pence to the decimal system. Crown Lynn released a cup and saucer to mark the changeover, which was a major event for the country.**

(Below) **Willow pattern.**

99

Through the late 1960s and 1970s the Titian factory made a large range of coffee ware, kitchenware and vases under the Crown Lynn banner. This honey-coloured glaze was very much in demand.

CHEMISTRY / Through the 1960s Crown Lynn put a huge amount of effort into research and development. Larry Moore, who started work in 1967, remembers around 20 laboratory staff working in four separate laboratories. 'There was always something new on the horizon. You were always developing something. There was always a challenge.' All these tests created an enormous number of experimental plates, which – officially at any rate – staff were not allowed to take home. A few were sold in the seconds shop, but hundreds were thrown into a huge pit and smashed. Sometimes, though, a plate with cryptic pencil marks on the back turns up in the second-hand market. The numbers and letters relate to the volume number and page number of the laboratory's glaze 'recipe books', in which the composition of all glazes was carefully noted. Moore recalls hiring a holiday bach at Tairua in the Coromandel: 'I went to the cupboards and there was all this Crown Lynn pottery. It all had my handwriting underneath. I was thinking, where did these come from? They were lab samples. Never did figure that out.'

ECHO AND PONUI / 'Hippies influence design for latest dinnerware . . . Crown Lynn has turned on with the flower children,' announced the company magazine in December 1969 as the new Echo design was unveiled. Designer Mark Cleverley was quoted as saying he had drawn the design 'with reckless abandon' to appeal to the younger set. This all-over flower pattern proved to be one of the best sellers of all time, well outlasting the flower-power era in which it was launched. It was still on the market in 1979.

Crown Lynn was very proud of this new release. The technical experts broke new ground in their efforts to turn Cleverley's vision into tangible cups and plates. First, the flower pattern was printed onto the white body in pure black, using the Murray Curvex

A jug and covered butter dish in the Apollo design. The butter dish, made for the export market, is shaped to fit American-style butter pats.

102

machine. Next, black was rolled onto the rim of the plates by hand to eliminate any sign of the white body. Then the plates were sprayed with a semi-opaque glaze in an ochre colour, which allowed the black all-over flower pattern to shine through. Finally, the backs of the plates were sprayed with a dark brown and the finished plate was put through the kiln.

At the same time as Echo was released, Crown Lynn introduced Ponui, described as being in 'cool contrast' to Echo. Naming these two designs was an interesting process. As usual, various suggestions were bandied around. Echo was so named because it was a repeat – the same flower pattern was repeated over and over again. Alan Topham named Ponui after an enjoyable boat trip to Ponui Island near Auckland. The rich blue-green glaze reminded him of the sea. For an export line, though, the name wasn't a huge success – Australians and Americans struggled with the pronunciation.

THE APOLLO RANGE / The fluted Apollo range was introduced in late 1969. One of Crown Lynn's most consistent sellers in plain white, it was also sold in plain colours and decorated with various lithographs.

The range was developed while Alan Topham was in Canada, where everyone was agog at the news that the first men had landed on the moon. Topham had no hesitation in naming the new range after the Apollo 11 spacecraft – though its designer Mark Cleverley always thought the name had something to do with fluted Grecian columns.

A MASTERPIECE / One of Crown Lynn's most impressive products came about almost by

From top: Ponui and Echo.

(Opposite page) The backstamp on this plaque describes the native flora and fauna it depicts. Only a few other plaques were marked 'Presented to the Rt Hon. J. R. Marshall on the occasion of his official opening of Ceramic House and presentation to Crown Lynn of the Trade Promotion Council's export award. March 1969'.

accident. The dark-brown plaque embossed in burnished gold had a real New Zealand flavour with detailed Maori designs, and flora and fauna including kiwi, kowhai and pohutukawa.

The inspiration came from a drawing that David Jenkin worked on for months, adding to it as time allowed. To make the plaque truly exceptional, the design was embossed into the plaque rather than simply printed on. An immaculate clay original was carved by Tam Mitchell, then a plaster of Paris mould made from it. The plaques were then made one by one. Liquid clay was poured into the mould and left until it consolidated. The plaque – still damp and very fragile – was taken from the mould and dried. Alan Topham describes the rest of the process: 'You fettle the edge – clean it all nicely – then do the first firing. The first firing is white. Then put on the brown glaze, then fire it again, then the gold is rolled onto the raised surface and any bits of gold that are in the wrong place are cleaned off.' Then it was fired again. Made from temperamental high-quality porcelain, many plaques sagged out of shape during firing. Some unglazed plaques found their way out of the factory, thus there are a few oddities in existence. Some are brown with no gold, others are pale apricot or pale green. It is likely that only about 300 of the full-size plaques were

made to be presented to important clients and suppliers in New Zealand and overseas. Crown Lynn also made a smaller wall plaque (which doubled as a teapot stand) in the same design.

CERAMIC HOUSE AND THE FIRST COMPUTER / Citing the need to keep track of a growing export market, in 1969 Crown Lynn brought one of the first computers into New Zealand. The machine was programmed and operated by six staff and occupied most of the ground floor of Ceramic House (built in 1969 in New Lynn), the new headquarters of Crown Lynn's parent company. Architect Neville Price explained that the vast plate-glass windows on the ground floor were designed to allow the public to view the new computer. 'It cost so much, and is a fascinating and complex machine, why shouldn't everyone see it at work?' In line with a government directive that computers should be put to maximum use, the machine was to be made available to other firms.

THE EXPORT AWARD / In March 1969 Deputy Prime Minister Jack Marshall presented Crown Lynn with an award for outstanding effort in the export field. While officially opening Ceramic House he was presented with the commemorative plaque designed by David Jenkin. By now Crown

Lynn was exporting two and a half million pieces a year to Australia, Canada, Malaysia and South Africa. The factory's 500 staff were producing 230,000 pieces a week – nearly 12 million pieces a year. At this time Crown Lynn was one of the many New Zealand firms which had built up hugely successful overseas markets. In the 1967–1968 period there was a 77 per cent increase in exports. The New Zealand economy was booming and the government actively promoted exports, offering financial incentives and practical help for companies that wanted to expand. Only about four export awards were presented each year, and previous recipients included Fisher and Paykel and Hamilton Jet. By the end of the 1960s Crown Lynn was making substantial profits. Throughout the decade the company had improved quality and increased production. At the company's twenty-fifth anniversary celebrations in 1969, Tom Clark noted that 'only nine years ago Crown Lynn was a dirty word'. Now he had 60 per cent of the domestic market and exports were soaring.

70s

LEVELLING OFF
1970–1979

107

In the early 1970s Crown Lynn reached its highest-ever output. Each week the factory turned out over 300,000 pieces, totalling 15 million a year. It used over 8000 tonnes of clay a year, most of it from Kerikeri, Mt Somers and Huntly. Approximately half the production was sold in New Zealand, and half exported. During this period the factory was at full capacity and the kilns were never turned off except for maintenance at Christmas. All this activity was extremely taxing on the families involved. It was routine for management staff to work 10- or 11-hour days. More often than not there was also work on Saturdays, and there was travel,

travel and more travel in the endless pursuit of export markets. John Heap recalled, 'It never occurred to you to challenge the hours, it was just part of the territory.' And there were compensations. Crown Lynn was a happy place to work at, and there was always time to have a few beers and a laugh with the guys. The wives, however, weren't always so enamoured of Crown Lynn. Alan Topham's wife Betty remembers sometimes insisting that her husband stayed home with his young family on Saturday morning.

Through the 1960s and 1970s Crown Lynn was bursting with energy. There was constant demand for new shapes, new designs and new colours. But no matter how exciting a new shape or pattern was, if it couldn't be made with Crown Lynn's machinery it was no use to anybody. The designers and marketers constantly pushed for innovation and the chemists and mould makers soon

(Opposite page) **These popular 1970s cups were designed in-house.**

From top: Truscan and Expo 70.

learned that there was no such word as impossible – if a job couldn't be done one way, they had to suggest an alternative. Through the 1970s the company bought new decorating equipment and developed new techniques that greatly expanded the range of decorative styles. Always, though, innovation was tempered with tradition. Householders liked to see fresh designs, but they tended to favour the familiar. If they had used a floral dinner set in pastel tones for decades, they were not likely to suddenly rush out and buy a replacement with a dramatic geometric pattern in dazzling colours. In the factory there were often lengthy debates over the pros and cons of new patterns, but in the end Alan Topham chose the designs that he thought would sell. Some designs, for example Mark Cleverley's Truscan, were ahead of their time. Cleverley felt that Truscan was one of his best, and the marketing team were also keen, but it didn't sell especially well.

THE EXPO 70 RANGE / In an effort to increase sales in Japan, the New Zealand Meat Board set up a pavilion at Expo 70 in Osaka, Japan, and Crown Lynn was commissioned to provide dinnerware for its restaurant. Mark Cleverley and David Jenkin created a simple white-on-white design, emulating the 'geyser' theme of the pavilion itself. For six months the restaurant dinnerware was put through commercial dishwashers up to six times a day – and still came out looking very presentable. Along with the white plates, Crown Lynn created a rectangular 'Legend of Maui' service platter designed by David Jenkin, and a white fishhook-shaped chopstick rest. The Chateau range of ovenware, glazed in dark green, was also created for the expo. Made at the Titian factory, and glazed in brown, this range was later sold commercially in New Zealand. The white plates and the platter were never sold on the domestic market, though some found their way into New Zealand through the seconds shops.

NGAKURA WARE / After the limited success of Wharetana Ware in the early 1950s, Crown Lynn did not endeavour to break into the souvenir market until the launch of the very similar Ngakura Ware in 1970. Designed at Crown Lynn and manufactured by the Crown Lynn-owned Luke Adams factory in Christchurch, the range was named in honour of long-serving employee Ringi Ngakuru. It included a pin box, cigarette box and ashtray, and a small rectangular bowl. No one is sure why the name of the ware was Ngakura, not Ngakuru.

THE LAMPSHADES / In the 1960s and early 1970s Crown Lynn manufactured a few ceramic lampshades in a simple bell shape. They were hand-made and never produced in commercial quantities; although a 1970 photograph of the new Crown Lynn shop in Auckland's 246 building shows 'the very popular ceramic shades' hanging in the foreground.

THE ASIAN VENTURES / By the late 1960s Tom Clark was pondering a burgeoning wage bill and a chronic shortage of reliable workers. He became convinced that if Crown Lynn was to prosper, he had to establish a factory in a low-wage, low-cost Third World country. This was not a popular view at the time. There was a strongly held belief that industry should be kept in New Zealand to protect New Zealanders' jobs. Crown Lynn's first attempt at an Asian venture was an investigation into setting up a factory in Korea, but the Koreans withdrew before it got started. In 1968 Crown Lynn negotiated a management contract with Starlite Ceramics in Singapore. Managed by John Heap, the factory made porcelain dinnerware which was exported to Indonesia, Canada and the Middle East.

However, an overly optimistic order book, run-down equipment and civil unrest combined to force Crown Lynn to abandon the venture and in 1971 the factory closed down.

The Mayon Ceramics factory in the Philippines was the only Asian venture that met with any success. The project was classed as overseas aid under the Colombo Plan and was partially funded by the New Zealand Government. The building project was supervised by Chris Harvey from Crown Lynn, who stayed on to manage the factory. Clay scientist Harry Jones was responsible for finding raw materials. In 1972 he travelled almost the length and breadth of the country with a Filipino counterpart, checking clay deposits and testing samples. The new factory, a 30 per cent joint venture with Filipino interests, was modelled closely on Crown Lynn's.

Harry Jones lived in the Philippines for months while the factory was being set up, and visited at least once a year for the next 10 years. The tropical climate had its drawbacks, as did the wildlife: 'Once I came across a snake. In the drier. A poisonous

Mayon dinnerware.

Green Bamboo   **Impala**   **Louisa**

(Opposite page) **Hand-painted planters from the Mayon factory.**

snake. That shot me back. And the weather. Stinking hot. We drank cases and cases of Coke and Fanta.' But there were good times, too; when Jones' wife Jean came to stay, the pair enjoyed the street markets and other aspects of the exotic culture.

As part of the aid agreement Crown Lynn was required to teach technical skills to their joint-venture partners, so 16 Filipino technicians spent several months at Crown Lynn learning about mould making, glazes and clays. The factory was managed by New Zealanders for five years, then Filipinos took over, with Crown Lynn continuing to provide technical support. David Jenkin designed a range of hand-painted planters and dinnerware for Mayon and spent time at the new factory training the hand decorators. Most Mayon products were sold in the United States and Australia, though some planters were also sold in New Zealand.

Although it showed more promise than other Asian ventures, Mayon was never a huge success. President Ferdinand Marcos'

wife Imelda reneged on assurances that Mayon would be the sole large ceramics factory in the Philippines, and soon other factories sprang up and the country was flooded with cheap locally produced china. Crown Lynn had expected to sell much of Mayon's output on the local market and this simply didn't happen. After Mayon had limped along for nearly a decade Tom Clark reluctantly came to the conclusion that there were 'too many holes in the bucket'. Corruption was rife and the company was paying substantial backhanders to get raw materials into the factory and to get the finished product out. Clark estimated that about 25 per cent of production 'disappeared' before it could be sold. Crown Lynn refused to put any more money into the venture and it closed in 1982.

The failure of the overseas ventures was a huge disappointment to Tom Clark. Decades later, he said that Crown Lynn would have had a good chance of survival if he had been able to set up a successful factory overseas.

Today, the English and Japanese fine bone china companies make a large percentage of their ware in low-cost countries such as Bangladesh, China, Indonesia and Sri Lanka.

ROYAL GRAFTON / By the late 1960s Tom Clark was ready to move beyond the earthenware and vitrified hotel ware manufactured at Crown Lynn. He wanted to make fine bone china, in the belief that this would open up more markets in the United Kingdom and the United States. Rather than starting from scratch in New Zealand, it seemed simpler to buy an English going concern. In 1968 a Crown Lynn team arrived in Stoke-on-Trent to buy the Royal Grafton factory – furtively, so as to avoid the attention of the 'big boys' such as Royal Doulton and Wedgwood, which both coveted it and certainly didn't want colonial upstarts buying into their patch. John Cowdery, who had been managing Luke Adams Pottery in Christchurch, returned to England to run the new venture. In 1974 Royal Grafton employed

This souvenir of the Punakaiki pancake rocks is typical of the bone china from Crown Lynn's Royal Grafton factory in England. Royal Grafton dinner sets sold well in New Zealand.

Crown Lynn in the late 1970s and early 1980s. The pieces now found in New Zealand were presumably sold in Crown Lynn seconds shops.

THE CROWN LYNN WORKFORCE / Apart from a strike in the early 1950s, Crown Lynn was remarkably free of 'union trouble' in an era when unions had power and were not afraid to use it. Tom Clark's relationship with his workforce was not always harmonious but he was liked and respected by most of his staff. He paid regular visits to the factory floor and knew most of his workers by name. In return most felt part of the Crown Lynn team and worked hard. On the negative side Clark was always trying to cut costs, and in a labour-intensive business that meant low wages. Union delegate Jo Crawford described pay rates in the 1970s and 1980s as 'absolutely criminal', but Clark argued that unless costs were kept down, Crown Lynn's products would not sell.

Productivity incentives were part of Crown Lynn life. As early as 1950 there was a bonus system, calculated on both quantity and quality of output. Tom Clark was also a strong promoter of time-and-motion studies. In the early 1960s specialist consultants reviewed each job to eliminate time-wasting actions and processes. John Heap, who helped introduce the system, said that as a result staff numbers in the moulding department were halved and overtime was cut. Surplus staff were moved to other parts of the factory. The regime was not always well received by the workforce, who felt coerced into working harder. Slower workers were not actually

250 people and manufactured dinnerware, coffee sets, breakfast sets and ornamental pieces. Its products were sold in England, Europe, Scandinavia, the United States, Canada, South Africa, New Zealand and Australia.

Unfortunately, fine bone china went out of fashion and Royal Grafton's prosperity didn't last. Crown Lynn sold the factory in the mid-1980s and a few years later it was demolished.

The occasional piece of Crown Lynn china marked 'Grafton Ironstone, Exclusively for Devaz, Greece, Made in NZ' turns up in New Zealand. This backstamp exists because a Royal Grafton customer, the Greek company Devaz, distributed earthenware made by sacked, but many resigned under pressure from supervisors whose bonuses depended on the productivity of their entire team. Another incentive scheme was introduced to combat post-payday absenteeism. Pay was distributed on Thursdays, and workers who turned up on Friday went into a draw for a household gadget, usually a toaster or an iron.

In another, far less popular move, a clocking-in system was instigated – arrive at work a minute late and your pay was docked. Tea breaks, too, were rigidly kept to 10 minutes, and it took some workers almost that long to get from the back of the factory to the tearoom. Chris Harvey, who was a cadet in the early 1970s, remembered factory manager Fred Hoffman's 10-minute breaks: 'A minute before the 10 minutes was up the bell went. You had to put your cup down, race back out the door, and Fred would be down at the making machine and as the main bell rang he would put the machines on, and the guys would be running . . .' At the time this policy was accepted with good humour by most of the workforce, but later it caused some contention.

Most Crown Lynn workers belonged to the Labourers Union, with two delegates representing the entire staff. On the whole the staff-management relationship was co-operative, but Tom Clark was not a supporter of the traditional union system, which he described as 'bosses in this corner, workers in that, and you slug it out'. In 1969 he said he feared that New Zealand had imported industrial troublemakers from Britain and Australia, and by 1971, a period of high employment and labour shortages, he was

These plates in the Last Wave design carry the Devaz mark. This design was advertised in New Zealand in 1981.

critical of New Zealand workers, who 'don't give a damn, just drift from one place to another'. He praised Third World workers who, he said, got stuck in, the way New Zealanders used to. In reality Crown Lynn's staff were no sluggards. Mark Cleverley recalled visiting a pottery on the west coast of Ireland in the 1970s. 'He (the manager) said he had been through the pottery at Crown Lynn and he said, "Oh, those Maori girls, I wish I had some of them here, they can really work."'

All in all Crown Lynn was a pretty good place to work. There were plenty of parties, sports teams and informal socialising. The cafeteria was friendly and cheerful – even if

the air was thick with cigarette smoke. And there were always cafeteria manager Harry Cheeseman's scones to look forward to. Above all, for most of its life, Crown Lynn was seen as a family business, and that atmosphere was fostered by Tom Clark, who mixed easily with his workforce. And, as one new employee discovered, he could forgive an honest blunder. Young Larry Moore, newly in charge of a laboratory, was told that no one was allowed in without permission. One day he returned from a break to find two men in suits in his laboratory. 'I said, "Excuse me, gentlemen, this is a restricted area and you aren't allowed in here." And one gentleman in

a very deep voice says to me, "Do you know who I am?" And I replied, "I don't give a f**k who you are, you're not allowed in here."' That afternoon, Moore was asked to report to his boss, Harry Jones: 'He said, "I hear you are really looking after the security of the laboratory, but please don't tell the managing director of Crown Lynn to f**k off." Tom thought it was a great joke. I was absolutely petrified at what I had done.'

THE FORMA RANGE / Advertised under the heading 'Nothing makes dish washing easier', the Forma range was launched in 1973. Developed first for the American market, it also sold well in New Zealand. Patterns ranged from the black and brown Toledo to the dazzling pink and yellow floral Gay. Forma's flat-bottomed, straight-sided shape was described as 'extremely modern in its concept' and the cup and saucer were self-docking – a ridge on the bottom of the cup fitted neatly into a groove in the saucer. The Forma range was created for the American retailers J. C. Penney, which in effect wanted Crown Lynn to copy a successful American range and sell it at a cheaper price. The shapes were designed on the run. David Jenkin and Mark Cleverley scurried back and forth between their drawing boards and visiting buyers from J.C. Penney until the Americans were satisfied. They were uncomfortable copying other people's work, but if that was what the client wanted, then so be it. At any rate the Forma range didn't sell well in the United States, largely, it was said, because there was a 'great void' between the price Crown Lynn wanted and the price J. C. Penney

(Right) **Forma Toledo.**

(Opposite page) **Forma serving platters. From top: Charmaine and Gay.**

was prepared to pay. For the Crown Lynn technicians the Forma range was a challenge. The designs called for big bold-coloured transfers to be applied to biscuitware, then topped with a clear glaze. Crown Lynn had never done that before and new techniques had to be developed. That was a frustrating process, recalled Larry Moore. The transfers held little pockets of air which would pop out through the glaze and destroy the finish when the plates were fired.

As with any new product, the designers tried to create a fresh new look for the Forma range. Any designs that showed promise were put onto a test run of around 100 plates. Many were rejected at that stage and the samples sent down to the seconds shop. One design created a sensation and was snapped up by customers in minutes. Unfortunately, the design was technically difficult and was never put into production.

THE SOCIAL SCENE / In the 1970s, when Crown Lynn was at its largest, the factory covered a substantial area with the pottery, the brick and pipe works, carpenters, painters and an engineering workshop all on site. The place was like a small village and gossip was rife, recalled long-time employee Tom Hodgson: 'You heard lots of rumours in a place like Crown Lynn. Some were made up and some were real. You didn't know what to believe – but you believed them at the time, you passed them on.' An informal arrangement allowed staff members to sell their produce around the complex – smoked mullet, kumara and meat packs were all hawked around the factory at various times. One kiln operator grew his own tobacco and dried it behind the kiln. Crown Lynn, like most industries, subsidised the staff social club which held regular events, including a Christmas party with 'Tiny' Percival as Santa distributing gifts to the children. Social club barbecues on the local beaches were popular, and there was a host of sports teams, with hard-fought rugby matches between Crown Lynn and the brickworks.

(Above) **A recruitment advertisement that appeared in the** *Western Leader.*

(Opposite page) **A Crown lynn hand painter.**

for celebration. Staff parties in Crown Lynn's heyday in the 1960s and 1970s were legendary. There were stories of marathon drinking sessions, especially on the last day before the Christmas shutdown. Each department was allocated a certain amount of alcohol and money to buy food, and there were parties all over the factory. Chris Harvey remembered his first party as a young cadet when temporary fences were put around production areas to prevent accidents. There were Polynesian dances and late at night a staff member did handbrake turns in his Mini on the new flooring in the cafeteria. The salaried staff, who often worked long hours with no extra pay, were given their own Christmas party.

A few of the younger members overdid the celebrations. There was a full bar and beautiful food, recalled one, but the party was always on a week night and everyone was expected to be at work on time the next morning. He always got up and went to work – 'But two years in a row they found me asleep on a workbench.' In the late 1970s the parties got out of hand, with too many gatecrashers and too much alcohol. A certain amount of licentious behaviour had always been tolerated – passionate embraces in private corners were common on break-up day – but the story goes that a brazen sexual act in the bus-stop outside the factory tipped the balance and the big company parties came to an end.

THE PACIFIC ISLANDERS / Back in 1963 Lyall Smith from the Taranaki Farmers Co-operative told assembled Crown Lynn dealers that the New Zealand worker was the finest

Especially during the early 1970s, when staff were in short supply, Crown Lynn was promoted as a fun, 'happening' place to work. An advertisement in the local paper reported on a highly successful car rally and noted that there was a free dance and social at the Mt Eden-Mt Roskill hall in Dominion Road.

Tom Clark respected Crown Lynn's hard-working employees and believed that any achievement was a good reason for a party. 'I would walk through the factory and see all those girls with their heads down, go go go. They were wonderful people, wonderful people.' A new export market, a new range of china, an increase in production, more tariff protection – there was always a cause in the world. 'He is more intelligent, more adaptable. Show him a problem, and he'll show you how to solve it.' Problematically, by the 1970s, this paragon was in extremely short supply, and businesses in Auckland were desperate for staff. The economy was prospering and there weren't enough workers to fill Crown Lynn's 500–600 jobs. The company began running four-hour twilight shifts and some departments also ran night shifts, but they were less successful, largely due to problems with quality. To recruit more staff, full-page advertisements were placed in local newspapers promising subsidised buses, staff buying privileges, an active social programme and company-supplied smocks to protect clothing. There was a crèche for the children of working parents and, in a technological breakthrough, wages could be paid direct into an account at the post office or bank.

Despite these incentives Crown Lynn still struggled to find workers. Throughout the 1960s young rural Maori moving to Auckland had helped fill the labour gap, but by the end of the decade the company was looking to the Pacific Islands. Crown Lynn representative Ed McCaffery went to Tonga and Samoa and signed up workers on the spot. The company paid the recruits' fares to New Zealand and found them accommodation in Auckland, then docked their wages until costs were repaid. Tom Hodgson, who managed the plate-making section in the 1970s, recalled that there was never a problem with finding more staff. 'I would say to the operators, have you got somebody at home, and a few days later

I'd have more workers.' Government work permits were never an issue either, largely because of Crown Lynn's status as an exporter. However, in some circles hiring immigrants was not a popular move and Crown Lynn was accused of importing cheap labour to keep wages down. By 1976 the large Pacific Island population in Auckland had attracted political attention and the government initiated the notorious 'dawn raids'. Police swooped on Polynesian families at first light, arresting and deporting anyone who was in the country illegally.

At Crown Lynn the introduction of large numbers of immigrants created its own problems. For many New Zealand workers it was the first time they had come across anyone who did not speak English. The new immigrants looked very different too – some

of the women wore their floral dresses and woven hats decked with flowers over the top of their Crown Lynn uniforms. To compound assimilation problems, immigrants from the different islands did not always get on, and there were occasional fights on the factory floor. Whenever possible, different ethnic groups were separated into different departments. The people who applied cup handles, for example, were almost all Maori, with factory manager Fred Hoffman explaining that 'the natural rhythm and finger-ability of Maoris makes them twice as efficient as anyone else in the delicate process of making a cup handle look as though it had grown with the rest of the cup'. There were also a number of European immigrants at the factory, most notably Tony Rakich, a Yugoslav who loaded and

unloaded the drying machines for years. After a couple of decades of walking backwards, pulling tall, heavy trolleys on wheels, he was conspicuously pigeon-toed.

EQUAL PAY AND EQUAL OPPORTUNITY /
Before the Human Rights Act was passed in 1971 and equal pay legislation in 1977, employers could pay male and female workers different rates for doing exactly the same job, and could openly discriminate between men and women. In 1960 an advertisement for a 'tall married man' for night work at the kilns caused a great deal of good-natured comment in the newspapers. Fred Hoffman explained that he needed a tall man to unpack high shelves, and he should be married because 'a single person will take odd evenings off to go to dances or the theatre, but a married man can't afford to'. Thirteen years later, in 1973, the company wanted new staff to get the latest factory extension into production. It promised bigger earnings for women as the 'first stage of equal pay puts high take-home pay in the grasp of all'. In 1977 design competition winner Mary Broadbent was described as a 'mother of two and former commercial artist'.

IMPORT RESTRICTIONS BACK ON THE AGENDA / In the early 1970s Crown Lynn

was again in the thick of the controversy over import restrictions. Tom Clark took out full-page newspaper advertisements maintaining that Crown Lynn deserved protection because it was employing New Zealand workers to make a quality product from New Zealand raw materials. The debate was heated, with importers saying that Crown Lynn had unfairly influenced the government. Clark's response was blunt: 'To say we have pressured the Government is an insult to the Minister of Trade and Industries, Mr Freer, who is quite capable of making up his own mind without outside help.' Crown Lynn was in a favoured position because it had a large workforce, it exported much of its product and it used mainly New Zealand materials. Whenever there was talk of removing import licensing, Tom Clark would pull out his trump card – he would head off to Wellington and say, 'Do you want to put 600 people out of work?' But despite his posturing Clark was well aware that import restrictions would not last for ever. As early as 1971 he said publicly that he was '100 per cent in favour' of the relaxation of import controls, but he wanted the change to be gradual.

CHANGING TIMES / For decades kiln chimneys belching smoke were accepted as part of the West Auckland industrial scene.

In 1960 the pipeworks was still salt glazing sewer pipes, pouring out salt-laden smoke which rusted corrugated iron roofs and corroded workers' cars. By the early 1970s the environmental movement forced a change. The year before the Clean Air Act was introduced in 1974, Crown Lynn began to install equipment to clean up the emissions from its chimneys. In December 1973 the company demonstrated a new 'smokeless' incinerator to ratepayers and ecological groups.

THE 'EARTHY' LOOK / After a trip to Canada and the United States in the late 1960s, David Jenkin noted that bold, earthy patterns were popular: 'There appears to be a corresponding decline in interest in the more delicate Japanese designs.' It was only a matter of time before the ubiquitous Autumn Splendour was finally knocked off its perch – though it was not discontinued until 1977. Through the 1970s New Zealanders began to recoil from industrialisation and mass production. Art potters set up workshops throughout the country, creating hand-made ware in muted browns and greys. In response Crown Lynn released the Chateau Craftware range, as close to a hand-potted look as the factory could achieve. The range was made at the Titian works, taken over by Crown Lynn in the late 1960s. The craft look extended

119

(Above) **This sugar bowl and cream jug set is a deliberate imitation of the work of studio potters.**

(Opposite page) **A Bellamy's dinner plate.**

new trend. Television cooks David Halls and Peter Hudson chose Rusticana – 'an attractive dark brown decoration with a speckled glaze' – for their studio kitchen, described in 1977 as the envy of any housewife. The Forma (1973) and Earthstone (1977) ranges also had their share of muted brown designs. Crown Lynn Technical Ceramics put out the Nobs range of brown door knobs – in contrast to the white and floral designs previously available. To cater for the growing band of craft potters, in 1978 Crown Lynn set up Western Potters Supplies next to the factory. The shop was staffed by working potters who sold clays and glazes and dispensed advice to aspiring amateurs.

THE BELLAMY'S DINNERWARE / Even the New Zealand Parliament jumped on the brown bandwagon. In 1977 Parliament Buildings were redecorated in autumn colours, with brown woodwork and brown drapes and carpets. In Bellamy's restaurant the white and gold tableware was deemed unsuitable for the new décor and Crown Lynn worked with a Parliamentary subcommittee to produce a new design. The committee included fashion-conscious MP Whetu Tirikatene-Sullivan, who gave her immediate approval to David Jenkin's Maori fish hook motif in shades of brown. Bellamy's ware was never released onto the open market, though a few items may have been sold through the seconds shop. The brown dinnerware did not last long. The next government wanted a change back to 'fine dining' and imported china from Noritake in Japan.

beyond dinnerware. In keeping with the new style, the Crown Lynn shop in the 246 shopping centre was finished in timber, quarry tiles, ebony brick and seagrass wallpaper. In another nod to the natural look, the Stoneware series – Rusticana, Sahara, Radiance, Focus and Tosca – was released in 1973, and in 1976 the Pioneer range was marketed as earthy, homely and appealing. It was snapped up by buyers keen to feature the rustic colonial look in their homes, and in Australia it proved to be one of Crown Lynn's most successful lines ever.

By the end of the decade the majority of New Zealand households had embraced the

COFFEE AND CIGARETTES / During the 1960s and 1970s it was considered sophisticated to sit in a brown-carpeted lounge serving coffee – usually powdered instant – from a matching coffee set. These sets usually consisted of a coffee pot, cream jug, sugar bowl and six straight-sided coffee cans with saucers, or, for the more daring, saucerless mugs. A cappuccino was made by mixing double strength instant coffee with hot milk. The coffee pot became a fashion item and appeared in ever more exaggerated shapes, sizes and patterns. Later, the matching coffee set gave way to a collection of brown coffee mugs suspended from a wooden mug tree on the kitchen bench. In some ways the coffee mug symbolised the rebellious mood of the young. It went hand in hand with protesting against the Vietnam War, smoking cannabis and wearing wrap-around skirts. Grandmothers sipped tea from matching bone china cups and saucers in floral patterns; the young people sat about in a cigarette haze drinking coffee from thick brown mugs. Gathering round the communal ashtray for a cigarette – or something stronger – was very much part of the coffee ritual. Over time ashtrays, too, became décor items, growing ever larger and more ornate. Crown Lynn produced a series of ashtrays emblazoned with company logos which were given away as freely as other promotional ware such as mugs.

CROWN LYNN HAS ARRIVED / By the mid-1970s Crown Lynn was an integral part of New Zealand life. Most households ate off Crown Lynn, even if they still kept their Royal Doulton for 'best'. You could scarcely open a magazine

This towering coffee pot is part of a set designed by Mark Cleverley and made at the Crown Lynn-owned Luke Adams factory. The entire set, which included a jug, sugar bowl and tall mug with the same distinctive rounded handle, was finished in dark brown or aquamarine. The coffee set was exhibited at the Design Centre in London. It first appeared in New Zealand in 1971 at the Crown Lynn shop at 246 in Auckland. Crown Lynn also produced several smaller but nonetheless exotic coffee pots such as the one on the facing page.

without seeing something made by Crown Lynn. High-profile chef Des Britten flipped his favourite omelette piperade onto Crown Lynn, and a *New Zealand Woman's Weekly* series on tasty dishes made from leftovers featured baked cheese savoury served on Crown Lynn Apollo. An advertisement for New Zealand Railways showed a happy family eating from Crown Lynn ware as the world whizzed by outside the window.

THE AIRLINES / In 1973 Air New Zealand discontinued its Crown Lynn order for the turquoise and brown china used by first-class passengers, but Crown Lynn was given the large and lucrative contract for economy-class dinnerware. Mark Cleverley and David Jenkin designed new shapes including oval dishes, plates and 'dinky little' cups that stacked neatly one on top of another. Glazed in dark brown with a black koru emblem, they were considered very stylish. They had to be shaped to sit neatly together on an airline tray and to fit into the narrow shelves in the air hostess' trolleys. The first order alone was for 330,000 pieces. Crown Lynn also made crockery for the Australian airline Qantas and for British Airways. A new machine, known as a ram press, speeded up the manufacture of oval dishes and other irregular shapes, which had previously been slipcast.

A NAME CHANGE – AND A YACHT / During the late 1960s and early 1970s Crown Lynn bought out Luke Adams and Titian potteries, followed by a steady stream of businesses in other sectors including transport, manufacturing

and engineering. An important purchase was Gibsons and Paterson, an importer and distributor of hotel and hospital ware. These purchases meant that the business was no longer reliant on the success of Crown Lynn pottery. It is likely that by now Tom Clark and his advisers could see that manufacturing china in New Zealand did not have a long-term future. By 1973 there were about 50 subsidiary companies in the Crown Lynn group, which listed assets of $10.5 million and a turnover of $27.8 million. Tom Clark's brother, Malcolm, retired and Tom became managing director

of the whole conglomerate. He moved to Ceramic House, the new company headquarters, and became less involved with the day-to-day running of Crown Lynn. Alan Topham was appointed general manager. By this time, though, Clark was no longer completely focused on the business. Weary of the pressures of running a large and still growing enterprise, he turned to sailing for relaxation. Despite its newfound size and influence the underlying company was still known by its original name Consolidated Brick & Pipe Investments Ltd. Clark decided – with considerable

(Above) Tom Clark with Peter, Pippa and Sarah-Jane Blake aboard *Lion New Zealand*.

(Opposite page) The Earthstone range. Anti-clockwise from top left: Polynesia, Sandown and Landscape. David Jenkin first created the Landscape design as an experimental piece to test the properties of the new in-glaze transfer technique. It caught the eye of the marketing team and became the best-selling design in the Earthstone range.

justification – that this sounded 'bloody Victorian' and in 1974 the company was renamed Ceramco. The new name remained low-profile until the company sponsored Peter Blake in the 1981 round-the-world Whitbread yacht race. Blake's boat was named Ceramco and, said Clark, 'I thought what a marvellous way to launch a new name on the New Zealand public ... it would get an enormous amount of publicity if we did any good. So I backed Peter Blake and he backed me and away we started. That was the beginning of Ceramco, really, the launching of that name behind a boat that went round the world.' This was also the beginning of a close professional and

personal association between Tom Clark and Peter Blake which lasted until Blake's death in December 2001.

THE CARBINE MODEL / In 1977 Crown Lynn reached new technical heights when the Auckland Carbine Club commissioned a 30-centimetre-high statuette of the famous racehorse Carbine, winner of the 1890 Melbourne Cup. It was modelled by Tam Mitchell, whose many hours of research included studying relics of the deceased animal. The mounted head was in the Auckland Museum and the hide of the 'great beast' was kept draped over a chair at the Ellerslie Racecourse. For several years a new statuette was presented each year to the winner of the Carbine Club Stakes at the Auckland Racing Club. It is not known how many of the statuettes still exist. Because of the risk of failure, three were made each year. One went to the racing club, and the others – if they survived firing and glazing – were scooped up by Crown Lynn staff. Crown Lynn also created ashtrays for the Carbine Club. They were decorated with a horseshoe imprint and a picture of the horse.

THE EARTHSTONE RANGE / In 1977 a new gas-fired kiln allowed the company to use a new 'resist' (or in-glaze) glazing technique. One of the first products of the new process was the Earthstone range, released first in North America in 1977. The five patterns, Autumn, Camille, Polynesia, Sandown and Landscape, were designed by David Jenkin. The thick silk-screen transfers in earthy brownish tones created a textured effect

(Above) **This service plate was made for the Beachcomber Hotel, a holiday resort in the Pacific Islands.**

(Opposite page) **A Ceramica ashtray.**

by melting into the glaze beneath as they passed through the kiln. Each transfer melted slightly differently, creating variations from plate to plate. A year after the North American release, Earthstone Ware went on sale in New Zealand, Australia and the Middle East. A television advertising campaign featured aerial shots of Mt Tarawera and the Waikato River.

EXPORTS GROW / Throughout the 1970s Crown Lynn continued to export much of its product and in 1979 predicted that Australia and the Pacific Islands would absorb one-third of the following year's production. There was monogrammed vitrified ware in a hospital in Adelaide and in league clubs in Sydney; at the luxury end of the market, two Tahitian hotels and a hotel in Hong Kong commissioned their own designs.

Australia was by far the biggest market but Crown Lynn ware was also sold in South Africa, the United States and Canada, as well as Fiji and Papua New Guinea. Tonga, Rarotonga, Western and American Samoa, the New Hebrides and New Caledonia were also possible markets. In the Middle East small orders had come from 32 retailers in the United Arab Emirates, Saudi Arabia and Kuwait. Mr Daulat A. Mirchandani, of Saeed & Daulat Ltd, visited Crown Lynn to sign up a sole agency agreement for Kuwait.

THE CERAMICA VASES / During the mid- to late 1970s Crown Lynn released a very successful range of vases backstamped 'Ceramica Greenstone'. Made at the Titian factory, the range included several vases

on the base of Crown Lynn ware. Keeping an enterprise of this size ticking over required a lot of organisation. Raw materials, including 14 different types of clay, lithographs and glazes, all had to be ordered in or made on site. The kilns burned round the clock for 50 weeks of the year and repairmen had no choice but to scurry across the top, rubber boot soles melting as they ran. More than once a shirt caught fire as repairs were made in the intense heat. As the factory grew ever larger, theft became a problem – not in any huge way – but staff had a tendency to pick up anything different or interesting. When the designers put a test run through the kilns, they picked it up immediately after it came out eight hours later, otherwise they might never see it again. Tom Clark once asked for a special order, a few swans glazed in black, and they were duly put through the kiln, never to be seen again; someone had picked them up before he could.

(Above) **Ceramica Greenstone vases.**

(Below) **Sunflower.**

and an ashtray, embossed with a distinctive leaf pattern in various shades of green, white and brown.

NEW MACHINERY / In the late 1970s Crown Lynn was still growing, introducing new technology and buying up-to-the-minute machinery from overseas. In 1977 a new kiln, two new banding machines and a new screen printer improved and speeded up the decoration process. An automatic bowl-making machine made it possible to mass-produce casserole dishes and other cookware. Lastly, three new backstamping machines were installed to speed up and sharpen the application of brand stamps

THE SUNDOWNER RANGE / In 1978 the Sundowner range was launched with the catchphrase 'Golden hues of dusk captured in Sundowner'. Finished in a golden brown, lightly speckled glaze, the range featured 25 new shapes. There was a full dinner set and a range of oven-to-table ware. The flatter plates, oval oven dishes, and casserole dishes with matching ramekins were promoted as 'Scandinavian-style'. The shape of the new casserole dish in particular proved very popular and, with various glazes and decorations, remained in production well into the 1980s.

COUNTRY FAIR / Yet another 1970s rustic-style product was the Country Fair range. Designed at Crown Lynn and manufactured at the Titian factory, the range included jugs, a butter dish, a cheese dish and lid, spice jars, an oil and vinegar jar, ramekins, a salt pig, coffee mugs and storage jars with ceramic lids. In this range variations on brown are the most common glazes, sometimes topped off with a greenish trickle glaze. 'Right now,' said sales manager Geoff Agnew, 'people want to dress up their kitchens and they want bright, attractive things to do it with.' In the following year other colours including white and a pale apricot decorated with sprays of flowers were also made in the Country Fair range.

A RETURN TO FLOWERS AND PASTELS / Towards the end of the 1970s tastes shifted again as New Zealanders yearned to escape from the mantle of gloomy brown. In 1977 Crown Lynn introduced the cheery Sun Series – Sunseeker and Sunflower. Both designs were American lithographs and were added to the replacement range.

AVONDALE AND CLIFTON WOOD / In 1977 the Avondale Selection, with new shapes for plates and cups and fresh floral decorations, was launched. It was named after the Avondale racecourse, 'only a saucer's throw' from the Crown Lynn factory. With names like Byways, 'a delicate tracery in green';

Country Fair.

A small sample of the different shapes and
designs from the Avondale and Clifton Wood
ranges.

(Left) Springflower.

(Below) The fiftieth birthday trinket box (left). The leftover boxes (right) were decorated with transfers and sold in Crown Lynn seconds shops in the early 1980s.

Savannah, 'daisies in green and white'; and Bramble, 'a circle of flowers coloured brown', its designs were supposed to evoke the countryside. Some patterns were imported lithographs; others were created by Crown Lynn designers. In 1978 the Avondale Summer Selection was added to the range, and in 1980 Avondale Selection giftware was introduced. The giftware range was added to through the 1980s and heavily promoted before Christmas. In retrospect the Avondale range appears to be a collection of unrelated designs. There were floral coffee sets and dinner sets, butter dishes with fruit designs, and tankards featuring antique buses and scenes from early Venice. A sister brand, the Clifton Wood Selection, was also produced a little later. It was retailed solely by Woolworths department stores.

In another bid to retain the traditional market, the Springflower range was also developed in the late 1970s. Promoted as 'bringing a breath of spring to your table . . . delicate floral designs to cater for more traditional tastes', its release was delayed until late 1979 after a fire damaged the factory's decorating department.

THE FIFTIETH BIRTHDAY / In 1978 Ceramco merged with Mason Industries to become one of New Zealand's largest industrial conglomerates, owning or partially owning more than 60 companies. In 1977 its sheer size had triggered a restructuring into three main divisions: engineering, distribution and services, and ceramics. In 1979, 50 years after the original Consolidated Brick & Pipe Investments Ltd was established, Ceramco

held a birthday party. Crown Lynn was commissioned to create an anniversary gift – a slipcast hexagonal lidded trinket box, decorated with sepia drawings of scenes from the company's early days. Designed by new head of design Tom Arnold, this gift and a leaflet detailing the company's history were presented to the company's 6300 shareholders.

131

80s

THE END OF A DREAM
1980–1989

Through most of the 1970s Crown Lynn appeared to be in good health. Production was steady, new machinery was installed and markets, both export and domestic, were satisfactory. But in the late 1970s the mood began to change. In 1978 a *Management Magazine* columnist recommended Ceramco shares, despite the fact that Crown Lynn, 'often a trouble spot for the group in the past', had again posted a disappointing result. At the fiftieth birthday celebrations, Tom Clark had been positive about the next 50 years, but he was talking about Ceramco as a whole. He believed that the conglomerate was now diverse enough to withstand temporary downturns – there would be enough profitable enterprises to support weak performers until things improved.

However, for Crown Lynn the future was anything but rosy. By the late 1970s New Zealand was entering a period of economic uncertainty: petrol prices were rising, interest rates had reached 15 per cent and would go higher, unemployment was on the increase and inflation was soaring out of control. New Zealand was borrowing heavily overseas. Since the mid-1970s import controls had been gradually eased, and more and more overseas-manufactured crockery was being sold in New Zealand. Government departments had been told to cut costs and inflation was biting into householders' pay packets. There was less buying power around, and Crown Lynn's domestic market suffered as a result. Exports also contracted as Asian manufacturers flooded world markets with cheap dinnerware, and England and the United States edged towards economic recession.

Ceramco's 1980 annual report advised that high domestic inflation and zero growth were making for difficult times at Crown Lynn. The company fought back by expanding its

(Opposite page) Florence.

product range, marketing more aggressively and cutting expenses. However, through the early 1980s the situation remained the same; Crown Lynn was losing ground in both domestic and export markets. The reasons for Crown Lynn's difficulties were not entirely confined to competition from imported crockery. The New Zealand market was almost saturated with Crown Lynn. It was a good-quality durable product and almost everyone who wanted a Crown Lynn dinner set already had one. New Zealand culture was changing. The days of formal dining had gone; most families gathered round the TV with their meals on their laps. This was a major shift from the traditions of the 1960s and 1970s, when young couples were given a Crown Lynn dinner set for a wedding present, then bought matching jugs, and cups and saucers off the rack. Domestic style, too, had changed. French Arcoroc glassware was in vogue and pottery was no longer the sole dinnerware option. Through the early 1980s the stock market soared and newly affluent investors were looking for novelty. International travel had become an affordable option and more consumers were aware of the range and variety of ware available overseas. Many of Crown Lynn's familiar products seemed increasingly out of step with modern trends.

THE SECONDS SHOPS / By the early 1980s Crown Lynn's warehouses held mountains of unsold biscuitware, which had accumulated during periods when sales had not kept pace with production. A retail manager, Stuart Spurr, was appointed to turn the surplus stock into money. First the biscuitware was

glazed and decorated – whenever possible with leftover lithographs. Then it had to be sold. To provide an outlet Crown Lynn built up its network of seconds shops from six to 21. Unlike other outlets these shops were controlled by the company, and the cash from over-the-counter sales went directly into its coffers. These were not strictly seconds shops, because they sold surplus stock as well as true seconds. Much of the surplus stock was not backstamped, while true seconds were. The surplus ware – apart from a mountain of saucers in discontinued shapes – quickly sold, and the company realised that the seconds shops were a consistent and important source of revenue. The factory began making new products to fill them. Most lines were successful, but an Easter promotion failed in spectacular fashion. Spurr got the factory to decorate 'millions and trillions' of egg cups in nursery rhyme designs including Humpty Dumpty and Little Bo Peep. He did a deal with Cadbury's to fill each egg cup with a little soft-centred chocolate egg – and they didn't sell. 'I still cringe at the thought of how many of those darn Easter eggs we ate,' he said. One of the most profitable seconds shops was in the low-income suburb of Glen Innes. The reason for its success was a mystery until a market survey revealed that heavily mortgaged Remuera residents were regular visitors.

A NEW GOVERNMENT / The country's economic troubles and the disastrous 1981 Springbok tour hastened the end of Robert Muldoon's reign as prime minister, and in

134 (Opposite page) A range of black mugs were made in the 1980s in a reaction to the fashion for black glass dinnerware.

1984 David Lange's Labour Government was elected. In a bid to cut inflation and reduce national debt, the new finance minister Roger Douglas opened up the country to foreign investors, floated the New Zealand dollar, deregulated the financial market and sold off many government-owned businesses. He also hastened the removal of import controls, a process which had begun in the mid-1970s. This was a crushing blow for New Zealand manufacturers, who had enjoyed 25 years of protection. The change affected many industries besides Crown Lynn, including makers of home appliances, vehicles, household goods and clothing. Through the 1980s controls were progressively lifted and by 1988 imports were pouring into the country.

For the New Zealand consumer this was a bonanza. Imported goods were cheaper, and, excitingly, there was more variety. Brightly coloured cotton T-shirts were astonishingly cheap, and sparkling new imported home appliances were stacked sky-high in discount shops. Chinese, Japanese and Thai – and even colourful Brazillian, Mexican and Italian – tableware could be picked up for a song. Too bad if it chipped and cracked; you could easily buy more. No longer were householders restricted to the familiar Crown Lynn designs or expensive English dinnerware. And the desire to 'buy New Zealand made' had all but evaporated. Twenty years before, in 1966, when J. Myers and Co opened their new Wellington showroom, Mr Myers proudly pointed out that everything on the tables – cutlery, dinnerware, tablecloths, napkins, place mats and glassware – was New Zealand made. Now buyers were hard pressed to find New Zealand products on the shelves.

During these economic upheavals, Ceramco was in the worst possible condition to deal with change. It had expanded almost out of control in the late 1970s and was now big and unwieldy. By 1984 it owned more than 60 businesses in New Zealand, Australia,

In 1981 Crown Lynn marketed a collection of souvenir items to mark the wedding of Prince Charles and Lady Diana Spencer. The range, which was sold through the seconds shops, included plates, at least two types of mug and a vase.

Canada, the United States, the Philippines and England. Ceramco had become a labyrinth, a complex interwoven web of companies – some fully owned, some joint ventures. Crown Lynn was only a very small part of Ceramco, which was a major shift from the days when Crown Lynn had been almost the sole focus of its parent company. Now Ceramco's managers had to consider the well-being of a large group of companies, not just a handful. And some of these companies were in direct competition with Crown Lynn. Hotel ware distributors Gibsons and Paterson were importing tableware for sale in New Zealand and Australia. New Zealand China Clays was exporting high-quality Matauri Bay china clay, an ingredient of Crown Lynn's products, to Asian china manufacturers. In June 1988, the year before Crown Lynn closed, 76 container-loads of clay were exported in one shipment alone – enough to make 50 million tea cups.

INTERNAL PROBLEMS / As early as 1982 a newsletter to Crown Lynn staff warned that there could be trouble ahead. There was a reasonable level of orders on hand, but the factory was finding it difficult to meet production targets. In a typical 1980s response to problems, the company laid off 40 staff and reorganised middle management. The next year Ceramco offered employees a chance to buy company shares at a discounted price with a three-year no-interest loan. As an effort to regain the team spirit of the 1960s at Crown Lynn, it was not entirely successful. The factory had lost its forward momentum and many workers felt less commitment to the job they were doing.

Crown Lynn's design department had lost its energy, too, and many of the products from this period seem pallid and lacking in substance. There was a surfeit of pink and grey, although this to some extent reflected the style of the day, when women's fashions ran to limp and shapeless pastel-coloured knitwear.

David Jenkin had retired in 1979 and not long afterwards Mark Cleverley also moved on. Crown Lynn hired a new head of design, Englishman Tom Arnold, and had a new marketing team. By then Geoff Honey had become general manager and Alan Topham was working for Ceramco, but not directly involved with Crown Lynn. Much later Topham said he had watched from the sidelines as Crown Lynn went downhill. 'It is my feeling that it got rather bland and the innovation disappeared. The designs looked wimpish to me, without making any real statement . . . I thought, Oh my goodness, that won't sell unless it's very, very cheap.' With so much cheap imported ware in the shops, Crown Lynn was under constant pressure to produce more designs, with faster turnaround and cheaper prices. This inevitably caused the company to gravitate towards safer middle-of-the-road styles. Crown Lynn viewed the New Zealand market as conservative, with a limited number of potential buyers adventurous enough to buy 'contentious' patterns. For example, the bold design Juliana was recognised for its excellence in the influential *Designscape* magazine, but was not for sale in New Zealand, being considered more suitable for sophisticated export markets. During the early 1980s the Crown Lynn design award, once a public relations mainstay, was quietly dropped.

Crown Lynn's internal problems compounded the difficulties it was facing in the newly competitive marketplace. Through the 1980s reports were sometimes positive, sometimes gloomy, but always there was an underlying suggestion that Crown Lynn was under siege. Restructuring, reorganisation, staff layoffs and cost-cutting were the order of the day.

TOM CLARK RETIRES / On 31 March 1984, at the age of 67, Tom Clark retired from his position as managing director of Ceramco, though he remained on the Board of Directors until 1993. By this time his involvement with Crown Lynn was minimal, but it was still a wrench to finally sever his connection with a company that had been central to his life for over 50 years. He recognised though that it was time to go after decades of intense and challenging work: 'Six o'clock in the morning and I'd be on the job. And I wouldn't get home till seven or eight at night, and this went on for years and years . . . But for me it proved not to be sustainable because I ran out of puff. By the time I retired I was about half the guy I had previously been.' To commemorate his retirement each staff member was given a mug embellished with a caricature of Tom Clark wearing a pirate's hat.

CONTEMPORARY CERAMICS / The imported crockery flowing into New Zealand had a substantial impact on Crown Lynn's market share. By the 1980s Crown Lynn was no longer the automatic choice for New Zealand households. In the *New Zealand Woman's Weekly* Crown Lynn seldom featured in photographs of Tui Flower's recipes for 'Summer Lunch' or 'Dad Takes Over the Cooking'. Magazines were peppered with advertisements for bone china and porcelain from Royal Doulton, Royal Albert, Noritake, Wedgwood and Arzberg. Even plastic Duraware had a higher profile. An expert was commissioned to tell Crown Lynn how to regain a decent share of the market. Among other recommendations, he said that the name Crown Lynn had negative connotations and the company should use a new brand.

Pink and grey designs were typical of the late 1970s and early 1980s. The mugs in this photograph were originally sold with saucers.

Thus, in 1980, Aurora Fine Stoneware was launched, without any reference to Crown Lynn. The first designs, Winter Wheat and Blue Lagoon, were promoted as 'an international collection of fine dinnerware … created in the classic English tradition'. This brand was followed by the Chelsea Collection by Contemporary Ceramics, again with advertising that made no reference to Crown Lynn. The range was principally pastel, often dominated by pink and grey. Mainly designed in-house, patterns included Ming, Dutch Iris, the popular Magnolia Moon and Elvenhood, a bolder design in brown and cream. Meanwhile, Crown Lynn also continued to promote its own brand, advertising pastel-coloured dinnerware with names like Spring Fair, Antique Rose and Wild Flowers – 'memories of a more gracious age'.

THE CORDON KITCHENWARE RANGE /
In 1980 the Cordon range of kitchenware, decorated with medieval kitchen scenes, was launched in Australia. The same ware was later sold in New Zealand under the Avondale Gourmet Selection brand. Around the same time Heritage 1800 Cookware was also marketed. It was decorated with sepia-toned scenes to co-ordinate with the Heritage dinnerware first produced in the late 1970s.

'LIFE AS NORMAL' /
Even as Crown Lynn struggled to come to grips with the realities of the marketplace, life at the factory continued as normal. A container-load of 24,000 pieces was shipped to Papua New Guinea, and in Tahiti the prestigious

Anuanua Hotel ordered its own distinctive vitrified dinnerware. Crown Lynn was selling well in Fiji and the Australian market remained buoyant. Nearly 1.8 million pieces of Crown Lynn were sold in Australia in 1985. The same year Minister of Broadcasting Jonathan Hunt chose the Spring Fair pattern for his ministerial residence in Wellington.

In 1986 the company celebrated 25 years of guided factory tours by hosting the Senior Citizens Club from Te Atatu. The group was welcomed by general manager Chris Harvey, and presented with a specially created plaque. There was even, in 1985, one last attempt to set up a factory overseas, a joint venture in China, but this, too, came to nothing.

Meanwhile, the financial reports continued to be discouraging. In June 1986 Chris Harvey reported that both the New Zealand and Australian markets were losing money.

(Opposite page) **Magnolia Moon.**

(Below) **Cordon Kitchenware.**

Crown Lynn planned to downsize operations in Australia, and 'get out of' Titian Fotteries by the end of the year.

NEW NURSERY WARE / Various forms of nursery ware were among the more successful lines through the 1980s. In 1987 Crown Lynn was advertising Norman Meredith's Nursery Tales designs. Characters from television shows such as *Paddington Bear*, *The Wombles* of Wimbledon Common and *Sesame Street* were popular through the 1970s and 1980s, and a range depicting Mickey Mouse and other Disney characters was advertised in 1982. One of the last nursery products was the Pirate Pete and Sally Anne series. Made in the mid-1980s they were designed so that the bowl, mug and plate stacked together to form a complete human figure. A Tootle Train set in the same style was also marketed in the 1980s.

THE COMPANY TAKEOVER / During the 1980s the New Zealand business scene was in a state of near-chaos. As the Labour Government reforms began to bite, long-established companies found themselves struggling to survive. In this state they were ripe for takeover and businessmen bought companies, closed down unprofitable operations and sold off disposable assets.

Among this group were Alan Gibbs and Charles Bidwill, who were major players in Ceramco by mid-1987. Ceramco became even larger through mergers with Atlas Majestic, which made home appliances, then with women's underwear manufacturer Bendon Industries. Twenty years later Alan Gibbs recalled that at the time all three companies badly needed restructuring to adapt to the changing economy. 'We endeavoured to restructure and to make a profit,' he said. 'That's what businessmen do.'

Once Gibbs and Bidwill had a majority shareholding they could take the lead in deciding what to do with Ceramco, and they began to shut down loss-making businesses and sell off assets. Among the first to go was the empty old Gardner and Clark brickworks between Rankin Street and Margan Avenue, which was demolished and the 15-hectare site sold. The Crown Lynn factory also came under scrutiny. Through the 1980s, and at times well before that, the factory had been at best marginally profitable, and had often lost money. In 1986 and early 1987, 170 more workers were laid off. Journalists were told that the layoffs were in response to declining sales, both in New Zealand and overseas.

THE OLD GUARD LEAVES / During its struggle to make ends meet Crown Lynn made a series of management reshuffles.

General managers were appointed, then just as quickly moved on when they were unable to meet revenue targets. Some long-serving staff members were sour about the merger with Bendon – 'from cups to cups' was the joke of the day – and they didn't appreciate the post-merger management structure. There was some consternation at the arrival of a female sales manager. Tom Clark had always referred to his managers as 'my boys', and some of the 'boys' did not take kindly to the new regime. At each successive reshuffle more of the old guard melted away. Many were burdened with disappointment and bitterness; they were leaving a company that they had put heart and soul into for decades. Long-serving ceramicist Harry Jones stayed until 1986, but chose to leave after a colleague warned that the business was in serious trouble. He was one of a group of about 14 middle managers who took redundancy at that time. Unhappy and disappointed, he didn't have the heart to say goodbye to his friend and boss Tom Clark.

THE SHAREMARKET CRASH / In October 1987 the New Zealand economy was dealt a shocking blow. Following the Wall Street sharemarket crash, New Zealand shares also tumbled, tipping the economy into recession. Investors lost their life savings,

140

(Above) Nova.

(Opposite page) Nouveau.

Crown Lynn was competing with equally boldly designed imported ware.

Throughout 1988 the news continued to be disappointing and in September a decorating kiln was shut down and 50 workers laid off. Mitchell told his staff that the company was manufacturing far more than it could sell. The warehouse was already overstocked by one million pieces, and at current production of 600,000 pieces a month the unsold stockpile would double in six months. By then the company's share of New Zealand's domestic tableware market was less than 20 per cent. This was a tense and difficult time. The workforce believed that the factory was heading for closure but Mitchell insisted it was to stay open. His desk, however, told a different story. The desks of previous general managers had been piled with files and papers; Mitchell's desk was empty apart from a phone, notepad and pen. Meanwhile, the layoffs continued. Some managers and administration staff who had access to sensitive information were given only minutes to clear out personal possessions and leave the premises.

THE END / When they returned to Crown Lynn after an anxious Christmas break, workers were told that if the company was to stay in business they would need to take a cut in overtime pay. Certain by this time that the factory was about to close, the union wanted better redundancy deals. The staff went on strike, but Mitchell made it very clear that unless they accepted the new pay rates and got back to work there was a very real chance that the factory would close for good. After two weeks on

people became nervous about spending money, businesses suffered and workers lost their jobs.

The year after the crash the new general manager Jonathan Mitchell told his staff that sales had slumped. To 'stay at the top' he hoped to increase sales in Australia as well as New Zealand. A boldly designed new range was introduced in an attempt to capture the mid-price market.

A LAST BURST OF CREATIVITY / Despite its dire financial position Crown Lynn found its design feet once again in 1987 and 1988. A new dinnerware shape, flatter with a wide rim, was introduced. The Modello Collection included the lively in-house designs Florence and Nouveau, produced in both black on white and white on black. This collection had a new sense of energy, but by this stage

the picket line the union appeared to have won. The workers settled for a four per cent pay rise and retained their overtime arrangements. By this time, though, both sides knew that the outcome was irrelevant; the factory was doomed.

At the end of April Crown Lynn held a huge sale to unload tonnes of surplus stock, and on 30 April the *Sunday Star* newspaper speculated that the factory was about to close. The story was illustrated with a photograph of an untidy pile of waste crockery, with an abandoned 'workers unite' placard in the foreground. The newspaper pointed out that it wasn't just Crown Lynn that was in trouble – the entire Ceramco conglomerate was struggling. Share values had dropped to one-third of their pre-crash high, and only Bendon and the Matauri Bay clay operation were making a decent profit. Jonathan Mitchell conceded that the factory needed about $2 million worth of improvements just to stay in business. Ceramco was not prepared to ever consider investing that sort of money, unless they got pay concessions from the workforce – concessions that union members were not prepared to make.

A few days later, on 5 May 1989, the last 220 staff were summoned to the company cafeteria and told that Crown Lynn was to close. The factory was shut down in stages, starting with the production departments at the back. Jo Crawford remembered this time vividly: 'You would be leaving at night from our department and you walked through the other departments to get to the main door, and some of these people would be coming along crying . . . As they closed down a department all the lights went out and the machinery was turned off. And it was kind of an eerie feeling. It seemed like every night – it didn't happen every night but it seemed very regular. Really sad. I knew that certain ones there would find it hard to get another job. Some had come straight out from the Islands and that was the only work they had known.'

At the time similar closures were happening all around New Zealand and Crown Lynn's laid-off workers faced a difficult future. In West Auckland alone they were competing against 6000 other registered unemployed. After 18 years at Crown Lynn, production worker Lavinia Wilkinson didn't mince words about her future prospects: 'It is going to be very hard to find employment and I don't have a clue what I am going to do . . . the conditions we were working in were terrible, but at least we had a job.' All the staff who were there at the last were given a railway mug with their name and years of service on it.

Even on the day that the closure was announced, the owners and managers of Crown Lynn voiced support for the government's reforms. Crown Lynn, said Jonathan Mitchell, had been 'caught up in a set of circumstances which makes this type of business no longer viable in New Zealand . . . We as a company are supportive of the liberalisation policies of the Government and have no argument with the overall aim of creating a leaner, more efficient economy.'

The workers, though, had a different view. Unionist Shirley Nusin told the newspapers that the government had 'opened the floodgates to all these imported things. We just couldn't cope unless we wanted to earn $20 a week like in the Philippines.' She blamed Ceramco for letting Crown Lynn's machinery slowly slide into disrepair. In an *Auckland Star* article a Labourers Union spokesman took a sideswipe at Charles Bidwill and Alan Gibbs who, he said, 'were more inclined to buy and sell companies than to manage and grow them'.

In September 1989 Crown Lynn's brand name and machinery were sold for an 'undisclosed sum' to GBH Porcelain in Malaysia. This announcement put paid to any hopes that the factory might be picked up by a local consortium and reopened. At the end of 1989 the Ceramco company newsletter made no mention of Crown Lynn. The front page was devoted to an article about Japanese customers visiting New Zealand China Clays at Matauri Bay and being treated to a feast of crayfish.

THE FINAL CLEAN-UP / Long-serving staff members Ray Machin and Chris Harvey organised the packing of the machinery ready to send to Malaysia. It filled over 40 shipping containers. The Malaysian company sent down six young men to help, but they were paid 'peanuts' said Machin. 'We used to feed them because they didn't have enough to live on.' The story goes that much of the Crown Lynn machinery is still sitting in unopened containers in Malaysia. Once anything saleable had been removed the factory was bulldozed. It took six months to clear the site. One ex-worker remembered her last visit to Crown Lynn. Walking across the flattened wasteland she found a pile of plates half buried in the ground. She gathered them up and took them home, sold some, kept the rest. Chemists Tom Clark Jnr and Frank Fitzpatrick were the last to leave. Not employed directly by Crown Lynn, they remained in their little laboratory, analysing clay samples and testing new glazes for overseas buyers, while the factory buildings were levelled all around them. When their laboratory was due to be demolished they packed up anything useful and moved to new premises only a stone's throw away.

WHY DID CROWN LYNN CLOSE? / The closure left many old Crown Lynn staff members saddened and bitter. They blamed the change in attitude and direction on Alan Gibbs and Charles Bidwill. Off the record, many said that the new owners were interested solely in selling up as much as they could and pocketing the money. There is some truth in this, but the reality is not so clear-cut. To be fair, most long-established New Zealand manufacturers were subjected to the same process at around this time. Those that survived came out the other end stronger, leaner and better able to prosper in the new competitive environment. But many went under.

The fundamental reason for Crown Lynn's demise was the fact that it had been built up in the artificial environment of import controls. Alan Gibbs believed that it had always been sheltered from economic reality, and would never have been a viable business in an open marketplace. 'At Crown Lynn people worked hard, and many were very skilful. But they were wasting their time. They had to face up to that. They were dedicated to making the best pottery they could, but they couldn't keep going after the government had taken away their protection.' To retain import protection Crown Lynn had to make the same range of products as would otherwise be imported.

'We made everything,' said Gibbs, 'but we were master of none.' In the end the large sprawling enterprise simply got too big and too complex for its own good. Closing the factory was expensive for shareholders, he said, and very painful for all concerned, but it was inevitable.

In retrospect Tom Clark, too, believed that artificial protection of New Zealand manufacturing was no longer an option. 'I was bending the thing round to the point where it was unsustainable.' It was quite apparent, he said, that Crown Lynn was going to go down the drain. 'It wasn't obvious to me, of course, but looking back it was quite obvious it couldn't be sustained.' Crown Lynn was especially vulnerable to the market reforms because of the type of product it made. The factory mass-produced mid-range and low-cost china, and this was the very market which the new Asian factories were aiming at – but the Asian product was much cheaper. By the time the factory closed, it cost more to make a Crown Lynn dinner set than it cost to buy an imported one from Noritake.

The lifting of import controls was only one of a number of circumstances which forced Crown Lynn's demise. There were also consistent problems with the marketing side of the business. John Goulter, who was executive director of the ceramics group of

Ceramco through the early 1980s, believed that Tom Clark tended to foster the manufacturing side, sometimes at the expense of marketing. 'Making things was the cornerstone of the Crown Lynn culture; it was more about making things than selling them.'

The overseas ventures, too, created more problems than had been anticipated. In keeping with an unspoken policy of looking after their own, there was a desire to staff overseas offices and factories with Crown Lynn people. However, not everyone wanted to uproot their families and live on the other side of the world, and not all of

those who were appointed had the experience to deal with the role. In the days before email and faxes, communications were difficult, said John Goulter. Crown Lynn executives were for ever crossing the world to visit 'the outposts of the empire, each with their own problems'. The Mayon Ceramics venture, which by 1980 was staggering to a halt, cast a pall over the whole enterprise. Its problems seemed insurmountable and Crown Lynn was spending its own money propping it up. During its last years Mayon was a blight on Crown Lynn. 'What was meant to be a flagship for Crown Lynn in the competitive world started to go sour.'

In the years after the closure Tom Clark spent many hours wondering if there was anything he could have done differently. He believed that Crown Lynn would have had a much better chance of survival if any of the overseas factories had succeeded. And he

wondered if he could have prevented the final demise of the factory by negotiating a deal with the new owners – but what sort of deal would have saved a business that was in such dire trouble? 'It wasn't the way I anticipated that Crown Lynn and Ceramco would finish up, but that's the way it did – which was a reality of the times. I could have made a bid myself for the company but I was conscious of Old Mother Time creeping up on me, and the fact that I wanted to live the rest of my life reasonably happy – not be under tremendous pressure.'

THE LEGACY / In New Zealand the large-scale ceramic industry died with Crown Lynn, and a huge bank of knowledge was lost or went overseas. A few smaller factories are still making a living, among them Temuka. In Auckland a direct descendant, Studio Ceramics, supplies niche markets with hand-crafted high-quality products under the control of former Crown Lynn general manager Chris Harvey. New Zealand's surviving potteries are small and quick to adapt to opportunities in the marketplace. Overseas many of the old English potteries have closed down. Among the survivors are Royal Doulton and Moorcroft; both have factories in low-cost developing countries.

Now, in New Lynn, there is little evidence that the Crown Lynn factory ever existed. There's a street called Crown Lynn Place, crowded with townhouses and apartments; there is a Clark Street and an Ambrico Place. The brick-maker Monier still has a base in the area and one solitary old brick-making kiln has been preserved.

This cup and saucer was made by Crown Lynn in 1987. It is reminiscent of the cheap dinnerware being imported from Asian countries at that time.

In their little laboratory in New Lynn, Tom Clark Jnr and Frank Fitzpatrick still develop new glazes and clay body – but all their work is for overseas companies. They still refer to technical notes put together by Crown Lynn staff in the 1960s and they still use the plate-making machine and mixing vats they salvaged from the Crown Lynn laboratory.

On the workbenches, filled with messy glaze mixes, sit old jugs and bowls that collectors now pay good money for. The chemists still use lithographs from some of the old Crown Lynn patterns in their test runs – a fragment of a bright orange poppy from the Avondale range lies among the piles of experimental plates.

Western Potters Supplies, set up in the 1970s to sell clay and glazes to home-based potters, also inherited some of Crown Lynn's equipment. Tom Hodgson, who worked there after Crown Lynn closed, recalled, 'I still feel like part of Crown Lynn and it is part of me … It was where I started my working life. It's a family company. It was a family. Fred Hoffman could tell you off, but he didn't hold any grudges . . . That's why I kept going back. It was family. It wasn't like that in other factories. Pottery was the centre of my life. Western Potters Supplies has still got some of the equipment that I walked up and down with when I was a youngster. It's got some of the racks there that were built even before I started. There are some tanks there that I remember being made. There's even one of the jack trucks, with a broken pedal that was broken when I was a boy. They are still using it and it still works.'

Today the name Crown Lynn lives on – in the Malaysian company GBH Crown Lynn. Owner Tony Goh says that his family business bought the assets of Crown Lynn in 1993, and now the Kuala Lumpur factory has a yearly output of three million pieces. Goh has a shop, the House of Crown Lynn, at Bangsar Shopping Centre in Kuala Lumpur. The upmarket fine porcelain and bone china is sold throughout Asia, and there are plans to export to Australia and perhaps even to New Zealand.

COLLECTING CROWN LYNN

As this book was being written, there was still plenty of Crown Lynn around at very reasonable prices. Some interesting pieces can still be picked up relatively cheaply from op shops and second-hand traders.

To get a clear idea of the enormous variety of items that were made by Crown Lynn, try frequenting the online auction sites. The busiest New Zealand site is TradeMe. Key in the words 'Crown Lynn' to get a list of every Crown Lynn item being offered for sale. It is also worth searching for other brands such as Kelston and Roydon as some sellers don't realise these were also made by Crown Lynn. Likewise, don't take it as gospel that every item listed for sale as Crown Lynn was actually made by Crown Lynn. Sometimes an item will be advertised and sold as Crown Lynn, and people with similar items will then believe that theirs are Crown Lynn – and so the myth is perpetuated. And remember that 'experts' don't necessarily get everything right. For example, there were a number of hand painters at Crown Lynn who decorated vases at various times in its history. One of Crown Lynn's hand painters once saw a vase which she remembered hand painting in the 1960s. She told the dealer her story and was told very firmly that she hadn't decorated the vase; Frank Carpay had. Items which are advertised as 'in the style of Frank Carpay' were not necessarily painted by Frank Carpay.

During your first visits to the online sites, it is a good idea to browse. Save your bidding until you know what you are looking for. After a few visits you will have a good idea of current values, though prices can vary wildly due to the auction process.

Although TradeMe is a good place to start, there are a number of other auction sites and private second-hand dealers' sites on which Crown Lynn is sold. It is worth Googling 'Crown Lynn' to see what turns up. Once you know what Crown Lynn looks like, and have an idea of what you would like to collect, it's time to study the list of backstamps at the back of this book and hit the shops. Remember that not all Crown Lynn items are backstamped Crown Lynn. Many have no backstamp at all, while others have one of the huge range of other marks which the company used at various times. Often the word 'British' is included – a reflection of the belief that the New Zealand housewife would prefer to buy English-made china. After a while you will get a 'feel' for Crown Lynn. Crown Lynn ware, especially the vitrified lines, is quite heavy. This is because it is made from a dense, non-porous body. The more porous the body, the lighter it is.

The glazes and shapes and patterns are distinctive, and there are a few identifying clues. For example, very few Crown Lynn items have flat bottoms – there is almost always an unglazed lip around the outside of the base of the item while the centre is glazed.

Since Crown Lynn made such a huge variety of different items and styles, it is a good idea to choose a theme. You may want to collect work by the design-award winners, or ashtrays, or monogrammed ware, or the 1980s pastels – or you may simply wish to complete a dinner set that you particularly like. If you choose a range that is already in demand, such as railway cups or work by Frank Carpay or Ernie Shufflebottom, prepare to pay premium prices. There are plenty of less ambitious choices, though, at reasonable prices. Whatever you decide on, be discerning. Think twice before you buy anything damaged or worn. The most collectible pieces are as close to new as possible. Unused boxed sets attract a premium. This can create a real conflict. Crown Lynn was made to be used. There is no reason why you shouldn't keep fresh flowers in your whiteware vase or eat off your Crown Lynn dinner set.

With Crown Lynn – and other New Zealand chinaware – increasing in popularity, you are up against some serious collectors. Even church shops and op shops are not the bonanzas they once were. But perseverance pays off. Most shops have a fast turnover so regular visits are a must. And check everything that 'might' be Crown Lynn. You will sometimes get a pleasant surprise. If you are looking for a specific item or range, it pays to shop around as prices vary hugely from shop to shop. If you find an item that you cannot identify with confidence, take it to a reputable dealer for confirmation.

And above all – enjoy! Whatever you choose to collect, appreciate its form and its design. Find out about its history. And take a moment to remember with respect the people who made it.

LOOKING AFTER YOUR CROWN LYNN / This section is not intended as a complete guide to the protection and restoration of china. If you have any doubts about what to do with your particular piece, ask an expert china restorer.

Museums keep their Crown Lynn collections locked in dimly lit air-conditioned storerooms. Obviously, this is not an option for everyday users and collectors, but there are some things that can be done to keep your Crown Lynn in good condition.

It is important to understand that Crown Lynn used many different decorating techniques and some are more susceptible to damage than others. The most vulnerable ware is decorated on-glaze – in other words the decoration has been applied on top of the glaze. The 1960s and 1970s design Topaz is a good example of an on-glaze decoration. If you hold a plate up against the light at exactly the right angle, you can see the small ridge which marks the outside edge of the lithograph. This means that the decoration is not protected by an outer layer of glaze. The decorations on the 1970s Forma range, on the other hand, are all underglaze, and are much more resistant to damage.

Next to breakage the biggest enemy of china is the domestic dishwasher. The combination of acid dishwashing powder and hot water causes patterns to fade and glazes to lose their sheen. Any gold highlights will soon disappear entirely. Even dinnerware which is labelled 'dishwasher safe' is not immune, although it will be slower to show signs of damage. By all means use your Crown Lynn in the kitchen, but remember that it will remain in top condition much longer if it is not subjected to the dishwasher. If your china is decorated with gold, always hand wash it.

China can also be damaged by everyday wear and tear, especially scraping with cutlery. Even simply cutting up food on your plate with a knife and fork can chip off the pattern. Once again on-glaze decorations are more vulnerable. Many Topaz plates are marred by visible scratches, where the coloured pattern has been scratched or chipped off, exposing the white body underneath.

Some plates are also susceptible to cutlery marks, especially from the lesser grades of stainless steel. These greyish pencil-like marks can be removed by rubbing very gently with a wet cloth dipped in a fine abrasive cleaner such as Jif. The wetter the cloth, the less likely you are to damage the glaze. The Crown Lynn Pine pattern is one that marks very easily.

Heat, too, will damage china, often beyond repair. Many a jug or gravy boat has been put in the oven to keep hot, and come out cracked and discoloured. Never put your Crown Lynn in the oven unless it is marked oven-to-table ware.

The fine cracks or fissures, known as crazing, which appear in glazes are very common in china, and Crown Lynn is no exception. Crazing occurs when the glaze shrinks faster than the clay body underneath it. It is not much of a problem unless the item is wet for long periods, for example, a saucer that has been placed under a houseplant, or a jug used to hold coloured liquid. The cracks develop black mildew or take on the colour of the liquid – this is generally impossible to remove.

STORAGE / The best place to store a china collection is on display. Provided it is in a secure earthquake-proof cabinet it should remain in good condition for decades. Plates can be hung on the wall, but both hanger and hook must be firm. If you cannot display all your china collection at one time, make sure it is stored in a sturdy container. A cardboard box is risky because cardboard flexes when it is moved. A safer option is a wooden or plastic box. Wrap each item well, preferably in bubble wrap or plain paper as newspaper ink can stain. Make sure your collection is kept dry. If you pack your china in a plastic storage box, drill ventilation holes in the sides and the lid to prevent condensation. Check your collection every few months to see that it is dry and in good shape. it is advisable to wash your collection and dry it thoroughly once a year.

RESTORATION AND REPAIR / Even the newest Crown Lynn ware is nearly 20 years old, and the oldest is approaching 60. Many items carry evidence of a long, hard life. Dinnerware can be stained from long use and inadequate cleaning, and the insides of vases are also often marked. If the item is not too badly discoloured, you may be able to bring it back to near pristine condition, though you will not be able to recreate the sheen of undamaged glaze. Try soaking the discoloured item in a nappy sanitising solution such as Napisan. If that doesn't work use a chlorine bleach such as Janola. After soaking it for around 12 hours, take your item out of the solution and rinse it thoroughly with clean water. At first it may show new marks where water has soaked into minute crazing cracks, but put it out in the sun for a few days and they should disappear. If any discolouration remains, try cleaning the article very gently with a wet cloth dipped in Jif. Do not use abrasive cleaners on valuable items.

If your piece of Crown Lynn is potentially valuable, do not attempt to clean or restore it yourself. Ask a professional china restorer for an assessment and quote, so that you can balance the cost of restoration against the cost of buying a replacement.

If the item is damaged beyond repair, consider making mosaics.

PRODUCT TIMELINE

Author's note: It would not be possible to list all the thousands of patterns which were produced by Crown Lynn during its 50-odd years of operation. Here I have attempted to list all major ranges and some of the more popular patterns. Dates, especially for the earlier products, are approximate. In the absence of reliable printed information I have had to take an educated guess in some instances.

1935 Acid-resistant tiles

1930s (late) Porcelain electrical components

1930s (late) First experimental jiggered household ware

1942 (approx) Chamber pots, plates, saucers, mixing bowls, casseroles, pudding basins, jugs, animal figurines, vases in plain glazes in blues and greens, hot-water bottles

1942 (approx) Unglazed blanks for Salisbury Ware

1942 (approx) Glazed blanks for Harwyn Pottery

1942 or 1943 (approx) American army ware. Cereal/soup bowls and handle-less mugs

1942 or 1943 (approx) Cups and saucers for New Zealand Railways and armed forces. No handles at first

1943 36-piece Paris utility dinner set in oatmeal colour, i.e., yellowish body

1944 Hot-water jugs

1945 (approx) Ambrico name used for the first time

1945 or 1946 Trickle-glaze vases

1944 or 1945 Jewellery

1946 (approx) Cigar jars, ginger jars, Seppelts bottles, shaving jugs

1948 Crown Lynn name used for the first time

1948 Good white clay body developed

1948 Introduction of new decorating techniques: gold, full-colour lithographs, monogrammed ware

1948 (approx) Toby jugs

1948 Ernie Shufflebottom hand-potted ware

1948 First white swans (previous to this, swans in other colours had been made). White swans were still being sold in 1973

1948 First matt white ware. Still being sold in early 1970s

1950 Empire games mug

1950s (early) First ovenware (casserole dish)

1949 or 1950 Wharetana Ware

1950s (early to mid-) Fancy Fayre Salad Ware, Wentworth Ware, decorated cake plates

1950s (early) Hand-made terracotta items

1950 Mirek Smisek's Bohemia Ware (until 1952)

1953 Frank Carpay's Handwerk series (until 1956)

1953 Coronation commemorative mugs

1955 Unglazed blanks again supplied to Salisbury

1950s (late) Cup handles made to stick properly

1959 First Murray Curvex automatic printing machine arrived. Not in full operation until early 1960s.

1959 Fleurette sold through DIC and Milne and Choyce. Still being sold in 1979

1959 First design competition. Winning designs Narvik and Reflections in production not long afterwards

1959 or 1960 The five favourites: Autumn Splendour, Golden Fall, Shasta Daisy, Green Bamboo and Fashion Rose. Autumn Splendour was still being sold in 1979

1960s (mid-) New lightweight vitrified coffee can and saucer. Designs included Mogambo, Blue Tango, Image, Allegro, Tacoma, Saraband and Bermuda

1961 (approx) Millefleurs door knobs and finger panels being made by porcelain department. Distributed in New Zealand and Australia

1961 (approx) Capri coloured ware

1963 Cook & Serve ware introduced. First designs were Narvik, Vision, Green Bamboo and Blue Tango

1963 Twenty new dinner set patterns and a new shape in dinnerware introduced. A total of 120 different dinnerware patterns in production

1963 Twelve new vase shapes in matt white and black, and a range of hand-turned television lamp bases

1964 Porcelain department making non-slip step tiles, car radio aerial insulators, ceramic door knobs and finger panels, soap dishes, towel rail ends, decorations for women's shoes, 'H' and 'C' tap knobs, reflective road markers and moulds for balloons and other items

1964 First spice jars

1964 Down Town series introduced

1964 and 1965 Nursery ware ranges on sale include Bunny, Wee Pets and Roydon Tiny Tots

1965 Topaz and Sapphire

1965 Dorothy Thorpe range: Pine, Palm Springs, Brocade, Monterey, Santa Barbara and Laguna. Ball-handled range sold in the United States. In 1966 a similar range without ball handles was sold in New Zealand

1965 Air New Zealand range introduced (brown on turquoise)

1965 Aztec, Mandalay, Sonata, Fabrique, Carousel and Coco. Storage jars and spice jars in Carousel and Coco introduced in 1966

1966 Blue Tango adapted for tourist hotels. Burgundy for the South Island, blue for the North Island

1967 Lynndale Range: Rose Red, Sierra Pine, Hacienda, Staccato, Capistrano

1967 Feminine Approach range of porcelain hardware, door knobs, keyhole covers, light-switch finger plates. Aztec, white and floral designs

1967 Decimal currency cup and saucer

1968 High Society range: Yucatan, Egmont. In 1969 Echo and Ponui were added to the range

1968 The new Shape twenty-5 range introduced. First products in the new shape included coffee sets in Novelle, Time Out and Carnaby

1968 Crown Lynn bought Titian Potteries. Thereafter most of Crown Lynn's whiteware and coffee ware was made at the Titian factory

1968 New range of porcelain bathroom hardware including toothbrush stands, towel-rail ends, toilet-paper holders. White with floral design.

1968 Willow pattern

1969 or 1970 Apollo range. Named after first moon landing, July 1969

1960s and 1970s Honey-coloured range of mugs in production at Titian factory by this time

1969 James Cook commemorative plate; only 2000 made

1970 The Expo 70 range: plates, cheese board, chopstick rests, a dark-green version of the Chateau Range

1970 Chateau Range released in New Zealand. Brown ovenware originally produced in dark green for the NZ Pavilion at Expo 70

1970 (approx) Ceramic lampshades

1970 Ngakura souvenir ware (made at Luke Adams pottery)

1973 Mayon Ceramics factory commissioned. Closed in 1982. Some Mayon items, mainly planters, sold in New Zealand during this period

1973 Forma Range: Toledo, Gay, Sahara, Radiance, Focus, Tosca, Rusticana

1973 Brown Rusticana glaze used on new shapes for Air New Zealand's new DC-10 planes. Casseroles made for Qantas and British Airways

1976 Pioneer range released. Brown glaze in an attempt to imitate art pottery look. Also produced in honey and white speckled glaze. Still being advertised in 1985

1976 (approx) Ceramica Greenstone range of vases and ashtrays

1977 Earthstone range introduced in the US. Launched in New Zealand in 1978

1977 Bellamy's new tableware

1977 Avondale Selection: Camelot, Savannah, Bramble and Byways were the first patterns. A wide variety of other shapes and patterns were introduced progressively under this brand through the late 1970s and early 1980s, which included Bon Cuisine kitchenware, Cottage Garden ovenware, cake plate and server, nursery novelties plaques, tankards and ashtrays, dinner sets, tea and coffee sets, jam and butter dishes

1977 Carbine racehorse model first created and presented

1977 Nobs range of brown door knobs from Technical Ceramics

1977 Sun Series: Sunseeker and Sunflower

1978 Ceramco's fiftieth birthday commemorative trinket box

1978 Sundowner range

1979 Springflower design. Cookware introduced in 1980

1979 Soup for One mug series advertised

Late 1970s to late 1980s Nursery ware included Nursery Tales range (advertised 1987), Wombles, Sesame St, Paddington Bear, Disney characters (advertised 1982), Pirate Pete and Sally Anne

1980 (approx) Heritage selection. Oven-to-table ware added in 1981

1980 Clifton Wood dinnerware range. Similar dinnerware to the Avondale Selection but exclusive to Woolworths. First designs included Granada, Serenade, Hampton and Somerset

1980 Cordon range of cookware introduced for Australian market. The same patterns released in New Zealand under Avondale Gourmet Selection brand

1980 Aurora Fine Stoneware range: Winter Wheat and Blue Lagoon

1980 Kiln collection: cups and saucers in earthy tones

1980 Heirloom collection: wall plaques featuring rustic scenes

1981 Charles and Diana royal wedding souvenir ware

1980s (early to mid-) Crown Lynn seconds shop network expanded. Much of the ware manufactured for these shops had no backstamp

1982–1983 Dinnerware and cookware in Wild Wheat, Heritage, Cherry Blossom, Amberglow advertised

1984–1985 Contemporary Ceramics Chelsea Collection advertised. Patterns included Magnolia Moon, Elvenhood, Quicksilver

1984–1985 Colonial Pottery advertised

1985 Modern Living range of dinner sets

1988 Modello Collection. Patterns included Florence, Nouveau

1989 (5 May) Crown Lynn factory closure announced

CROWN LYNN BACKSTAMPS

The backstamp is the mark on the back of a ceramic item which identifies its maker. Over its 40-year lifespan, Crown Lynn used many variations on the 'Crown Lynn' theme, but also used a wide range of other marks. I have recorded as many as I was able to find, but no doubt there are more. It has proven very difficult to date many of the backstamps with any certainty, so I have listed them in alphabetical order and given a date where possible.

I have provided photographs of the main styles and many of the more obscure ones. Space does not allow me to include a photograph of every backstamp which incorporated the words 'Crown Lynn' or 'Kelston' or 'Roydon'. Nor have I included all the variations on a theme – for example both Gigi and Gigi British were used.

Many Crown Lynn products bear no backstamp at all. Some of the early items are not marked, nor are many of the egg cups, small vases and other oddities. Ware made to be finished by other companies, such as the Salisbury Ware, was not marked. Items which were made especially to be sold in the seconds shops were not stamped, nor were many of the trial pieces which were also sold in the seconds shops or taken home by staff.

Many of the later Crown Lynn backstamps include the words 'Pat no.'. This indicates the pattern number, but was deliberately left as an abbreviation to give the impression that the design was patented.

In summary these are the main styles of backstamps used by Crown Lynn and its predecessor Ambrico.

154

THE FRACTIONATED MARKS (1940s and early 1950s).
These were used in trials to indicate the types of glazes and firing methods used. They were also used on commercially produced lines for a time.

IMPRESSED NUMBERS (from approx 1943 to 1960s).
These numbers were impressed on the base of moulded ware such as vases, jugs and swans. The numbering system started at 1 and continued through to about 900. Different sizes of the same shape were denoted by a dash and another single digit, e.g. 1, 2, or 3 depending on size.

STAMPED MARKS (1944 approx to 1989).
Various marks were stamped onto the base of the ware or applied as transfers. The first backstamp – a circled 'Made in NZ' – was initially used when Ambrico began exporting to Australia in the mid-1940s.

HAND-PAINTED MARKS (mainly 1950s) E.G. FRANK CARPAY'S 'HANDWERK' MARK.

THE FOUR-DIGIT NUMBERING SYSTEM (1964 to 1989).
In February 1964 a new four-digit numbering system was introduced. This number was often used in conjunction with the words 'MADE IN NEW ZEALAND'. Of the four digits, the first number indicated the type of ware and the second the material used. The last two digits were the design number. Some pre-1964 pieces were reissued using the new numbering system, thus it is possible to find two pieces which are the same shape but have different numbers.

Aero British
1950s (estimate)

AMBRICO
240v 1500w
Printed mark on electric jugs, 1946–1948

Apollo
Genuine
New Zealand
Ironstone
First used late 1969 or 1970

Ascot
1950s to 1960s

Ascot
1960s (estimate)

Ashleigh
Pat. No. 548
New Zealand
Other named patterns were stamped in a similar style.
1960s and 1970s

Aurora Fine Stoneware
First used 1980

The Avondale Selection (etc) Made in New Zealand
There are a number of variations to this stamp.
1977 to mid-1980s

Bunny by Crown Lynn
1950s and 1960s

Checkers by Crown Lynn
Many other pattern names were stamped in a similar style.
1960s and probably 1970s

Bamboo
1960s (estimate)

Calypso; Calypso British
1950s and early 1960s

Classic
1960s

Basil Brush
1970s

Candy dishwasher safe Made in New Zealand
Many other pattern names were stamped in a similar style.
Late 1970s to 1980s

Clifton Wood Selection
First used late 1970s

Blue Garland by Flair New Zealand
Late 1950s and 1960s

Capri British
Late 1950s and early 1960s (estimate)

Colour Glaze; Colour Glaze New Zealand (script)
1970s (estimate)

155

Blue Gingham (also Green Gingham)
From the early 1960s, many marks are in a similar style, incorporating the pattern name and number. From late 1950s or early 1960s

Capri by Crown Lynn
First used 1960 (estimate)

Colour Glaze (typeset circle)
Variations with the words 'Colour Glaze New Zealand' were sometimes printed in lines.
1970s and 1980s

Bohemia Hand Made (several different variations exist)
Used by Mirek Smisek.
1950 to 1952

Caribbean Ware
1960s

Contemporary Ceramics New Zealand
First used early 1980s

Bouquet
Late 1950s and 1960s

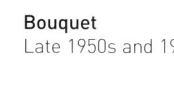

Ceramica Green Stone by Crown Lynn
First used mid- to late 1970s

Coronet
Late 1950s and early 1960s

Coronet symbol (no words)
1970s and 1980s (estimate)

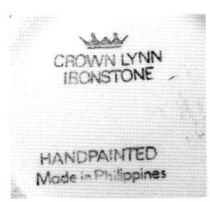
Crown Lynn Ironstone Handpainted Made in Philippines
1973 to 1982

Crown Lynn 'Stars Mark'
Used on first hand-decorated transfer ware.
1948 to 1950

Covent Garden British
Late 1950s and 1960s

Crown Lynn Microwave Cookware
1970s (estimate)

Crown Lynn New Zealand 'Tiki Mark'
Several variations occur including a green and yellow mark.
1945 to 1955 (estimate)

Crown Lynn (script)
1948 to 1955 (estimate)

Crown Lynn New Zealand (inside scroll)
Variations without the words 'New Zealand'.
1948 to 1955

Crown Lynn Vitrified
1948 to 1955 (estimate)

Crown Lynn stick-on label
1948 to 1955

Crown Lynn New Zealand
Similar marks with three stars. A larger version of this mark was introduced from 1960 (estimate).
From 1955

Crown Manor
Late 1950s and 1960s

Crown Lynn Earthstone
First used 1977

Crown Lynn New Zealand Super Vitrified Ware
From 1970

Dad Series
Late 1950s and 1960s

Crown Lynn for Barbara's
1970s or 1980s

Crown Lynn New Zealand Vitrified
1960s (estimate)

Devon Maid Ware
Late 1950s and 1960s

Crown Lynn Forma
First used 1973

Crown Lynn Philippines
1973 to 1982

Dishwasher Safe Detergent Proof Microwave Safe Made in New Zealand
1970s and 1980s (estimate)

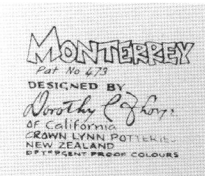

Dorothy Thorpe (series)
On tableware for both the American and New Zealand markets.
From 1965

Down Town
From 1964

Excella Ware
Late 1950s and 1960s

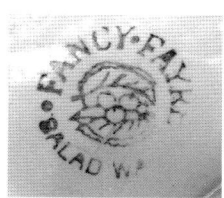

Fancy Fayre Salad Ware
Late 1940s and early 1950s

Ferndale British
Late 1950s and 1960s

Fiesta;
Fiesta by Kelston Potteries
Late 1950s and 1960s (estimate)

Fiesta Gaye
by Crown Lynn
Late 1950s and 1960s

Fiesta Ware
Late 1950s and 1960s

Flair Art Pottery New Zealand
Used on slipcast ornamental wares in matt white and pastel colours.
1960s (estimate)

Fleurette
Brereton Ware;
Fleurette
Late 1950s to 1970s

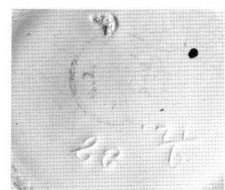

Fractionated marks
1940s and early 1950s

Gay Gold;
Gay Gold British
Late 1950s and 1960s

Genuine Ironstone;
Genuine Ironstone New Zealand
Late 1960s to 1970s or early 1980s

Gibsons and Paterson;
Gibpat (etc)
Gibsons and Paterson also imported ware for sale in New Zealand.
From late 1960s

Gigi;
Gigi British
1960s

Gina
Exclusive to United Stores
1960s (estimate)

Gold Lace
by Crown Lynn
Late 1950s and 1960s (estimate)

Goldline
Kelston Potteries
1960s

Golden Bouquet;
Golden Bouquet British
1960s (estimate)

Grafton Ironstone
Exclusively for Devaz Greece
Made in New Zealand
Late 1970s or early 1980s

Hand Crafted
by Crown Lynn
New Zealand
Made by hand potters (Daniel Steenstra) and hand decorated.
Mid-1950s

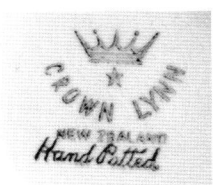

Hand Potted
Used with other marks, for example the Tiki. Used by Ernie Shufflebottom and possibly other hand potters.
1948 to 1950s (estimate)

Handwerk
Used by Frank Carpay.
1948 to 1950s (estimate)

Health Department Crown Lynn
1970s (estimate)

Heritage 1800 Crown Lynn Potteries
Late 1970s and early 1980s

**Horizon
Acid & Detergent Resistant Colours**
1960s (estimate)

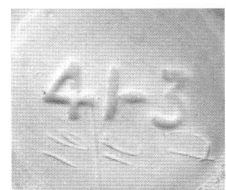

Impressed numbers
A few items, e.g. swans, jugs and vases, were still being numbered in this way in the 1960s and possibly 1970s.
From 1943 (estimate)

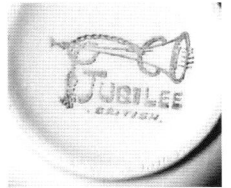

Jubilee British
Mid-1950s and 1960s

Kelston British
From 1962

Kelston Ceramics; Kelston Potteries New Zealand
1960s to 1980s (estimate)

Kelston Potteries Oak Folk
1970s and/or 1980s (estimate)

Kelston Ware British
1960s (estimate)

**Lido;
Lido British**
Late 1950s and 1960s

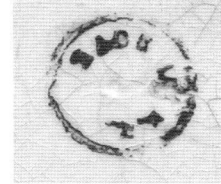

Lynndale Ware
There are many different patterns in this range.
From 1967

**Made in N.Z.
(inside circle)**
1943 to 1950 (estimate)

**Made in N.Z.
(with Crown Lynn coronet symbol)**
1970s and 1980s (estimate)

**Made in New Zealand
(inside circle)**
1943 to 1950 (estimate)

Made in New Zealand
Variations on this mark were widely used through the 1960s and 1970s

**Made in New Zealand
(inside circle; modern script)**
1970s

**Made in New Zealand
(plus four-digit number)**
This example was made at Titian Potteries after it was bought out by Crown Lynn.
1964 to 1989

**Made in New Zealand
(with design number)**
1980s

Nursery Novelties Avondale
After 1977

**Nursery Rhymes
by Kelston of
New Zealand**
1970s

Pussy Willow British
Late 1950s and/or 1960s

Rose Marie British
Late 1950s and 1960s
(estimate)

**Nursery Tales
by Crown Lynn
New Zealand**
Same style of stamp
used on other designs
as well.
1970s or 1980s
(estimate)

Recipe Collection
1970s (estimate)

**Royale
from the Quartet Range**
1980s

**Nursery Time
Kelston Potteries**
On same types
of ware as Wee Pets
British.
1960s

Regal Potteries
Late 1950s and 1960s

Roydon Pottery
1960s

**Oven Proof
Cook & Serve
by Crown Lynn**
From 1963

Regal Rose
1960s

**Roydon South Pacific;
South Pacific**
From mid-1960s

**Paddington & Co Ltd
Filmfair Ltd**
1970s (estimate)

**Rembrandt
by Crown Lynn**
1960s

Roydon Tiny Tots
1960s

Paris
Late 1950s and 1960s

Rigo by Giftime
Mid-1960s or early
1970s

**Roydon Wood Land
(also Roydon Harvest,
Roydon Tam-O-Shanter
etc)**
After 1967

Patio Rose British
1960s

Rosalie
1960s

Savoy
Late 1950s or
early 1960s

**Seagrass
Acid and Detergent
Proof**
Mid-1960s and/or early
1970s (estimate)

Sylvia Corvette
1960s and/or early
1970s

**Two by Two
Crown Lynn**
1960s to 1970s

Silver Maple
Mid-1960s and/or early
1970s

Sylvia Rose Kelston
1960s and/or early
1970s

Wee Pets British
Mid-1950s to mid-1960s

Sovereign British
Late 1950s and 1960s

Symphony British
1950s or early 1960s

Wentworth Ware
Stick-on label
also used.
Mid-1950s and
early 1960s

**Starline;
Starline British**
Late 1950s and 1960s

**Talahasse Distinctive
Tableware**
Mid-1960s or
early 1970s

**Wharetana
(stick-on label)**
Early 1950s

160

Sterling
1950s and possibly
1960s

**Teddy
by Crown Lynn
New Zealand**
1950s

**Wild Flowers
Crown Lynn
New Zealand**
A similar stamp was
also used earlier,
possibly in the 1950s.
Late 1970s or 1980s

**Sweet Sixteen
Distinctive Tableware**
Other pattern names
also in this style.
1960s and/or 1970s

**Tradition
Made in
New Zealand**
This example
is printed over an
Apollo stamp.
1970s

**Wild Rose
Made in New Zealand**
Late 1970s and 1980s
(estimate)

Sylvia Blossom Time
1960s (estimate)

Tudor
1950s (estimate)

**Wildlife;
Wildlife British**
1960s (estimate)

GLOSSARY

ASHET An oval plate.

BACKSTAMP The mark on the base of a ceramic item that identifies its manufacturer.

BALL MILL A mill that grinds up raw clay into a smooth paste.

BISCUITWARE Sometimes called bisque ware. Ware which has been fired but not yet glazed.

BLANK A pottery item which has been fired but not yet decorated.

BODY The clay mixture from which china is made. The body can have up to nine different components that are mixed together in mathematically determined proportions.

BONE CHINA A very white vitrified body of great translucency. It contains a high proportion of bone ash. Much of the very thin china ware made in England is bone china. Crown Lynn never made bone china but its English subsidiary Royal Grafton did.

CLAY Clay is formed by the decomposition of feldspar and in chemical terms is dehydrated silicate of aluminium. If mixed with water, clay becomes plastic and when dried it becomes hard. When a clay item is fired, it becomes harder still.

CRAZING A network of fine cracks that sometimes appear in the glaze after firing. Crazing occurs if the glaze contracts at a faster rate than the clay body. Glazes can continue to contract for years after the item is made, so crazing may take a long time to develop.

DRY-PRESSING A method used by Crown Lynn to make porcelain items. A metal die was filled with powdered clay and massive pressure applied to consolidate it. The item was then fired in a kiln.

EARTHENWARE A slightly porous type of body, of white or ivory colour. Most household ware produced by Crown Lynn was earthenware.

FELDSPAR A mineral containing potassium or sodium, as well as alumina and silica. It is mixed with other clays to make the clay body.

FETTLING Finishing the surface of a ceramic article before it is fired. The surface is smoothed with a tool, emery cloth or wet sponge. Flat articles are usually placed on a spinning wheel to speed up the process.

FILTER PRESS A filter press turns liquid clay into workable clay by pressing it through a series of filters to remove excess water.

FIRING The process of putting ware through a kiln. Ware is stacked high on trolleys and carefully separated by fireproof 'kiln furniture'. The trolleys spend up to two days going through the kiln. The ware must be heated up slowly and cooled slowly, so the temperature is highest in the centre of the kiln.

FLATWARE Saucers, plates and dishes are called flatware.

FRACTIONATED MARKS Some of Crown Lynn's products were marked on the base with various numbers and letters. These were experimental pieces, and the marks indicated the types of glaze and clay body used. Sometimes the date was also inscribed.

GLAZE Glaze is a form of glass, mixed with different metal oxides to create different colours. It is applied to biscuitware or greenware, then fired in a kiln at high temperatures. The glaze coats and protects the porous clay body and waterproofs it. Without glaze, water and impurities soak into the ware, discolouring and contaminating it.

GREENWARE Pottery items that have been shaped but not yet fired.

HAND MODELLING Shaping clay by hand, using simple tools.

HAND PAINTING The process of painting colours onto ware, usually with a paintbrush. Crown Lynn's Fleurette dinnerware pattern was hand painted.

HAND THROWING A lump of clay is placed on a potter's wheel (driven either by a foot pedal or electric motor) and the potter shapes it by hand as it spins.

IRONSTONE Another name for earthenware. A marketing term used by Crown Lynn.

JIGGERING A mechanised version of the potter's wheel. A measured block of clay is placed on a rotating mould and a template pressed down onto it, forming the clay into the shape of a cup, bowl or plate.

KILN A furnace in which clay items are fired. Usually tunnel shaped, kilns are made of heat-resistant material such as firebricks. The ware is loaded onto heat-resistant shelves on wheeled cars and moved slowly through the kiln for up to 48 hours. The centre part of the kiln is usually the hottest. At this point the ware becomes white-hot.

LINING AND BANDING Applying decorative lines or bands (wide lines) to ware, especially to plates, cups and bowls.

LITHOGRAPH A method of decorating china ware. The pattern, printed on special paper, is soaked in water and applied to the ware with a sponge or squeegee. The ware is then fired, causing the design to bond with the glaze. Lithographs are generally full colour and detailed. Graduations of colour in the same shade can be achieved.

161

MALKIN PRINT A type of printing process sometimes used at Crown Lynn. The popular Tam-O-Shanter pattern was a Malkin print.

MURRAY CURVEX A decorating machine used by Crown Lynn from the early 1960s until the factory closed. The later machines could print up to three colours onto flatware. They were much faster than previous methods of decoration.

ONCE-FIRED WARE Pottery which has gone through the kiln only once. Most of Crown Lynn's products went through the kiln at least twice – once as biscuitware, again after glazing, and often yet again to set the decorative transfers applied on top of the glaze. The Titian factory produced once-fired ware.

ON-GLAZE DECORATION A decoration (hand painted, transfer or print) applied on top of the glaze. This makes the decoration susceptible to damage from cutlery, dishwashers, etc.

PINDISH A small dish suitable for jam or butter.

PORCELAIN White, translucent, vitrified ceramic ware. Fired at a very high temperature, porcelain is harder and more durable than earthenware.

PROUTY KILN A type of kiln that protects the ware from direct contact with smoke and flames from the burning oil used for heating. First used by Crown Lynn in the late 1940s, the Prouty kiln greatly increased the range of decorating methods used.

PUG MILL A machine that mixes the clay body and removes any air bubbles.

REFRACTORY Made to withstand high temperatures. Refractory bricks were used to line furnaces, kilns and domestic fireplaces. The Kamo brickworks made refractory bricks.

SAGGAR A fireclay container inside which ware is fired. Saggars were used in the early days before the Prouty kiln was introduced.

SALT GLAZE A simple glaze created by sprinkling salt into the kiln at high temperatures. The salt volatilises, coating the ware in a plain brownish glaze. Most often used on sewer pipes and other utilitarian ware.

SGRAFFITO (SOMETIMES GRAFFITO) A pattern made by scratching through an outer layer of coloured glaze to expose the contrasting colour of the body beneath. Used to great effect by Mirek Smisek in his Bohemia Ware.

SILK-SCREEN TRANSFER A process used by Crown Lynn to decorate ware. A silk-screen transfer is not as detailed as a lithograph. There are usually blocks of colour, with no graduations in intensity of the shade.

SLIP Liquid clay used for slipcasting.

SLIPCASTING Irregularly shaped ware is made by this method. Liquid clay is poured into a plaster of Paris mould. The mould absorbs the water from the 'slip' and in about 30 minutes a crust of firm clay has formed around the inside of the mould. The surplus slip is poured away to leave the piece of pottery clinging to the surface of the mould. Most slipcast items were made in two or more pieces and often a mark can be seen where the pieces were joined. Most of Crown Lynn's slipcast items have a number impressed into the base.

TERRACOTTA A reddish clay body. Only used at Crown Lynn for a brief period in the early 1950s to make hand-made items.

THROWING The process of forming shapes by hand, from clay turning on a potter's wheel.

TRICKLE GLAZING An item is covered in a base glaze, then a second and sometimes third colour is trickled over the top.

TUBE MILL A tube mill mixes clay to the right consistency for use.

UNDERGLAZE DECORATION A decoration (hand-painted, transfer, lithograph or print) that is protected by a layer of clear glaze over the top. This makes the decoration much more resistant to knife marks and wear and tear.

VITRIFIED WARE A term referring to a clay body that, when fired at a high temperature, becomes vitrified porcelain. The water absorption characteristics are extremely low, resulting in a tough and durable product. Vitrified ware is usually produced for heavy-duty use in institutional and commercial catering. Vitrified ware is more expensive to manufacture than earthenware, and only a limited number of decorations can be used on it.

WHITE BODY A mixture of clay and other materials that turns white when fired in a kiln. Clay that looks white when dug out of the ground may turn red in the kiln as impurities such as iron change their chemical nature; clays that are dark grey or brown may turn perfectly white as impurities are burned out in the extreme heat of the kiln.

WHITEWARE Decorative ware glazed in white.

INDEX

Page numbers in **bold** indicate illustrations.

165

ACKNOWLEDGEMENTS

My thanks to those employees of Crown Lynn – or family members of Crown Lynn workers – who shared their stories with me so generously. They are (in no particular order) the late Sir Tom Clark, Lady Trish Clark, Alan Topham OBE and Betty Topham, Colin Leitch, Chris Harvey, Bebe Cowdery, Melva Ockleston, John Heap, Ray Machin and Eileen Machin, Harry Jones, Tom Clark Jnr, Frank Fitzpatrick, Ken Martin, Alan Gibbs, Larry Moore, Jack Mason, Tom Hodgson, John Goulter, Guthrie Stewart and Mary Stewart, Gina Bays, Donna Bird, Max Eyes, Ross Brewin, Stuart Spurr, Mark Cleverley, Wendy Steenstra Bloomfield, Dom Steenstra, Audrey Jenkin, Roy Gibson, Val Mitchell, Chris Curley, Connie Clark, Geoff Clark, Jo Crawford, Gordon Dryden and the late Ed McCaffery. I also had access to interviews with Mirek Smisek, Heather Shingle and Merle Phelps. Many of these people also lent me documents to copy and precious items to photograph.

On the research side, I owe an enormous debt to Gail Henry whose pioneering work *New Zealand Pottery, Commercial and Collectible* was an invaluable source of information, and to Douglas Lloyd Jenkins whose work on Frank Carpay and other designers was also immensely useful. My research was also greatly assisted by the librarians at the Waitakere Library and Information Services, the Auckland Museum Library, the Auckland Public Library, the Waitakere City Council archives and the Alexander Turnbull Library. My thanks go to Ron Lambert from the Taranaki Museum, Louis Le Vaillant and Finn McCahon-Jones from the Auckland Museum, and to the many antique dealers and collectors, in particular Jim Drummond, Helen Slater and Peter Joyce, who shared their knowledge with me.

My photographers Haruhiko Sameshima and Mark Adams of Studio La Gonda did a superb job, as did designer Athena Sommerfeld. My thanks also to my mentor Stephen Stratford and the Penguin team – Bernice Beachman, Louise Armstrong and Rebecca Lal – and to Renee Lang, who introduced me to Bernice.

As I was writing this book I became involved in an oral history project initiated by the Waitakere District Council. Robyn Mason and David Dromer from that project provided valuable support, and Mary Donald and I worked together on a number of interviews. Waitakere Mayor Bob Harvey and members of the West Auckland Historic Society were also helpful and supportive.

On the personal front, I appreciate the support of my partner George Irwin and my children Josh Monk and Emily Lummis. My sister Gill Sanders, my Aunt Mim Ringer, my cousins John Ringer and Lesley Madgwick, along with Lois Williams, Alison Lees, Jan Twentyman, Paula Jackson, Susan Wylie, Kitty Godwin and Annie Everson-Dawn – all in their own way helped me develop this concept and see it through to its final form.

Lastly, all credit to the second-hand shop owners, church shops, hospice shops and Salvation Army shops throughout the land who recycle cast-offs for future generations – and to those who donate to them.

This book was written with the support of the Ministry for Culture and Heritage History Research Trust Fund and the Mentor Programme run by the New Zealand Society of Authors (PEN NZ Inc) and sponsored by Creative New Zealand.

BIBLIOGRAPHY

Black, Adam and Charles, *The Story of the Willow Pattern Plate*. Aldine Press, London, 1975.

Ceramco Limited, *Ceramco Limited: A History 1929–79*. Auckland, 1979.

Hales, Olive, *Crown Lynn New Zealand: The First 30 Years*. Exhibition catalogue, Gisborne Museum and Arts Centre, 1983.

Hayes, Lorelei, *Waiaua to Kauri Cliffs: The Story of a Northland Sheep Station 1833–2003*. I and L Hayes, Kaeo, Northland, 2005.

Henry, Gail, *New Zealand Pottery: Commercial and Collectible*. Reed Books, Auckland, 1999.

King, Michael, *The Penguin History of New Zealand*. Penguin Books, Auckland, 2003.

Lambert, Gail, *The Clark Family History: Ms Author Athol Miller*. Privately published, Auckland, 1989.

Le Vaillant, Louis, 'Considering Frank Carpay', *Art New Zealand*, no 109, summer 2003/04.

Lloyd Jenkins, Douglas, *Frank Carpay*. Hawke's Bay Cultural Trust, Napier, 2002.

—*At Home: A Century of New Zealand Design*. Random House, Auckland, 2004.

Morley-Fletcher, Hugo, consultant ed., *Techniques of the World's Great Masters of Pottery and Ceramics*. Christie's, Oxford, 1984.

Raw, June, *Rail Tracks and Chimney Stacks*. West Auckland Historical Society, Henderson, Auckland, 1998.

Scott, Dick, *Fire on the Clay: The Pakeha Comes to West Auckland*. Southern Cross, Auckland, 1979.

Sperber, Hannah, 'On a Plate'. *North & South*, November 2005.

Williams, Howard, *Howard Williams' Pottery Workbook*. Beaux Arts, Auckland, 1974, revised edition 1977.

Tape-recorded interviews June 2004–January 2006
Alan Topham, Bebe Cowdery, Chris Harvey, Colin Leitch, Eileen Machin and Ray Machin, Gordon Dryden, Guthrie Stewart, Harry Jones, Heather Shingle, Jo Crawford, John Heap, Ken Martin, Larry Moore, Mark Cleverley, Mary Stewart, Melva Ockleston, Merle Phelps, Mirek Smisek, Sir Tom Clark, Tom Clark Jnr, Frank Fitzpatrick and Tom Hodgson. Supplementary notes by Harry Jones, July 2005. Handwritten notes by Jack Mason, February 2006.

Archived collections
Crown Lynn Potteries, Records 1959–87. Auckland War Memorial Museum Library. MS 98/34.

Gail Henry papers, 1924–1990, Auckland War Memorial Museum Library. MS 2006/4.

Scrapbook of newspaper clippings compiled by Elizabeth Hindmarsh. Auckland Library, 995.71 S43.

Waitakere Library and Information Services Local History Collection and J.T Diamond Collection, Henderson Public Library.

Periodicals and newspapers
Auckland Star: 18 July 1959, 20 October 1959, 22 July 1960, 1 September 1960, 24 January 1962, 5 May 1962, 16 May 1962, 17 May 1962, 21 May 1962, 30 August 1962, 14 April 1969, 31 December 1970.

Business Times: C8, October 1982.

Designscape: Te Aro: New Zealand Industrial Design Council, February 1969–October 1983.

Dominion: 24 July 1959, 15 August 1963, 1 October 1965.

Evening Post: 7 May 1959, 12 May 1959, 27 May 1959, 20 July 1959, 19 August 1959.

Hawera Star: 1 June 1960, 2 August 1966.

New Lynn News: 23 April 1959, 31 August 1959, 26 February 1969.

New Zealand Herald: 15 April 1959, 18 July 1959, 16 July 1963, 18 August 1965, 20 July 1959, 15 September 1959, 31 October 1959, 24 January 1964, 13 August 1964, 9 October 1969, 3 April 1968, 27 October 1972, 18 July 1979, 8 October 1982, 11 March 1987, 19 March 1987, 21 November 1987, 6 May 1989, 18 June 2005.

New Zealand Woman's Weekly: 30 March 1959, 17 August 1959, 21 November 1965, 26 February 1973,

8 November 1976, 19 November 1977, 12 November 1979, 19 May 1980, 7 July 1980, 3 November 1980, 24 November 1980, 15 December 1980, 23 November 1981, 30 November 1981, 29 November 1982, 13 December 1982, 7 November 1983, 28 November 1983, 5 November 1984, 4 November 1985, 1 December 1986.

New Zealand Listener: 20 September 1971, 30 June 1984, November 1987, 12 December 1987.

North Shore Times Advertiser: 15 July 1959.

Northern Advocate: 16 April 1961.

Otago Daily Times: 7 October 1960.

Retailer of NZ: 10 October 1969.

Sunday Star: 30 April 1989.

Taranaki Herald: 8 January 1960.

Thames Star: 15 February 1963, 30 July 1963.

The Chronicle (Levin): 27 March 1961.

The Standard: 29 July 1959.

Waikato Times: 18 July 1961, 20 July 1961, 13 August 1964.

West Auckland Historical Society Newsletter: 267, October 2004.

West Auckland Press: 8 February 1966.

Western Leader: 16 May 1967, 18 March 1969, 20 May 1971, 5 June 1973, 23 September 1973, 25 September 1973, 6 December 1973, 16 August 1986, 11 December 1986, 28 February 1989, 14 September 1989.

Unpublished material

Clark, Tom, letter to John Cowdery, 2 July 1946.

Crown Lynn/Ceramco publications held in private collections.

Ceramco Annual Reports, 1978, 1980, 1981, 1984, 1985; general managers monthly report, 14 July 1986.

'Ceramics', 1976 (undated), 1977 (Summer, Winter), 1978 (Summer, Autumn, Winter), 1980 (Summer), 1985 (April, August), 1986 (September), 1987 (December).

Newsletters to Crown Lynn staff, 1982, 1983, 1988, 1989.

'New Zealand Ceramics', 1963 (January, February, March, July, September, October, Christmas, Queen's visit edition), 1964 (July, September, December, Easter, Olympic Preview), 1965 (March, June, October, December), 1966 (July, September), 1967 (March, July, November), 1968 (May, July, December), 1969 (December), 1970 (April).

'Potters Pie', staff magazine, porcelain department Amalgamated Brick Pipe and Tile Co. Ltd, October 1948, Christmas 1948–1949.

Stevens, Liana, 'David Jenkin, Industrial Designer', Research Paper 603, 1997, Unitec, Unitec Library, Mt Albert, Auckland.

Tour arrangements for the visit of Her Majesty Queen Elizabeth II, Crown Lynn Potteries, New Lynn, Auckland, February 8th MCMKXIII (Crown Lynn Potteries, 1963).

Winstone, Lauren, 'Daniel Steenstra, Ceramist', Research Paper 604, 1998, Unitec, Unitec Library, Mt Albert, Auckland.

ILLUSTRATIONS

All photographs by Studio La Gonda unless otherwise stated, and all items photographed are from the author's collection unless otherwise stated.

Abbreviations

AWMM Auckland War Memorial Museum
AWMML Auckland War Memorial Museum Library
OHC Olive Hale Collection
CLP Crown Lynn Potteries Ltd
CEDAT purchased with funds from the Charles Edgar Disney Arts Trust
NZSPF purchased with funds from the New Zealand Specimen Purchase Fund

Page 4, AWMM, OHC, gift of CLP, K5819.
Page 8, AWMML, MS 98-34, photographer unknown.
Page 10, AWMM, clockwise from centre: NZSPF, K4673; NZSPF, K4674; OHC, gift of CLP, 1989, K5775; K4342; OHC, gift of CLP, K5776.
Page 12, Waitakere Libraries, J. T. Diamond Collection, 2003.6.PH.05437.
Page 14, cover of *Potters Pie*, private collection.
Page 16, AWMM, top: gift of CLP, K6308; right: K4570.
Page 18, photographer Val Monk.
Page 19, AWMM, clockwise from left front: gift of Mr and Mrs David Reynolds, K3753; 1995X1.720; gift of Mrs J Gillies, K4542; K3905.
Page 21, photographer unknown, courtesy of Gina Bays.
Page 22, AWMM, clockwise from lower left: gift of Mr and Mrs David Reynolds, K4582; K6375; NZSPF, K4837; NZSPF, K4927; OHC, gift of CLP, K6273.
Page 24, photographer unknown, courtesy of Bebe Cowdery.
Page 25, photographer unknown, courtesy of Bebe Cowdery.
Page 28, AWMM, clockwise from front: OHC, gift of CLP, K6209; CEDAT, 1997.44.4; K6686; OHC, gift of CLP, K6211.
Page 29, AWMM, clockwise from left: CEDAT, 1977.44.8; OHC, gift of CLP, K6214; K4331; CEDAT, 1997.44.6; NZSPF, K4659; NZSPF, K4660; NZSPF, K5042; NZSPF, K4664.
Page 34, private collection.
Page 36, AWMM, K3714.
Page 42, AWMM, left: gift of Mrs M Brothers, K6600; right: K3725.
Page 43, AWMM, clockwise from front: K3710; gift of Richard Wolfe, K3890; K3887; K3778.
Page 44, AWMM, gift of Mrs Jessie Hargreaves, top: 2002.80.2; bottom: 2002.80.1.
Pages 48–9, AWMM, from left: OHC, gift of CLP, K6030; CEDAT, K5481; OHC, gift of CLP, K6031; OHC, gift of CLP, K6024; OHC, gift of CLP, K6042.
Page 51, AWMM, CEDAT, K7140.
Page 52, AWMM, clockwise from front: K6392; K6393; gift of Mrs Sherry Reynolds, K4681.
Page 55, private collection.
Page 56, AWMML, Ms 98-34, photographer unknown.
Page 58, private collection.
Page 59, AWMM, all NZSPF, clockwise from front: K4661; K5252; K4902; K4903.
Page 61, photographer unknown, courtesy of Bebe Cowdery.
Page 62, AWMML, Sparrow 2884B, Sparrow Industrial Pictures.
Page 64, AWMML, MS 98-34, photographer unknown.
Page 71, AWMML, PH 89-6, Barry McKay Industrial Photography.
Page 72, private collection.
Page 73, private collection.
Page 75, AWMML, MS 98-34.
Page 77, advertisement, courtesy of *New Zealand Woman's Weekly*.

Page 83, AWMML, PH 89-6, Barry McKay Industrial Photography.
Page 84, private collection.
Page 87, AWMM, from front: OHC, gift of CLP, K6188; OHC, gift of CLP, K6189; NZSPF, K4863; NZSPF, K4683.
Page 98, advertisement from *New Zealand Weekly News*, courtesy of APN News and Media.
Page 105, private collection.
Page 116, advertisement, courtesy of *Western Leader*.
Page 117, photographer Ans Westra.
Page 118, AWMML, PH 89-6, Barry McKay Industrial Photography.
Page 124, photographer unknown, courtesy of Alan Sefton.
Page 126, private collection.
Page 147, brochure, courtesy of GBH Crown Lynn, Malaysia.
Page 157, Dad Series and Devon Maid Ware backstamps from *New Zealand Pottery, Commercial and Collectible* by Gail Henry, courtesy of Gail Henry.
Page 160, Paris backstamp, from *New Zealand Pottery, Commercial and Collectible* by Gail Henry, courtesy of Gail Henry.

In loving memory of my mother Mary Ringer, who introduced me to the pleasures of fossicking through second hand shops, and my father Bob Ringer, who reminded me more than once that 'the key to life is an enquiring mind'.

PENGUIN BOOKS

Published by the Penguin Group

Penguin Group (NZ), 67 Apollo Drive, Rosedale, Auckland 0632, New Zealand (a division of Pearson New Zealand Ltd)
Penguin Group (USA) Inc., 375 Hudson Street, New York, New York 10014, USA
Penguin Group (Canada), 90 Eglinton Avenue East, Suite 700, Toronto, Ontario, M4P 2Y3, Canada (a division of Pearson Penguin Canada Inc.
Penguin Books Ltd, 80 Strand, London, WC2R 0RL, England
Penguin Ireland, 25 St Stephen's Green, Dublin 2, Ireland (a division of Penguin Books Ltd)
Penguin Group (Australia), 250 Camberwell Road, Camberwell, Victoria 3124, Australia (a division of Pearson Australia Group Pty Ltd)
Penguin Books India Pvt Ltd, 11, Community Centre, Panchsheel Park, New Delhi – 110 017, India
Penguin Books (South Africa) (Pty) Ltd, 24 Sturdee Avenue, Rosebank, Johannesburg 2196, South Africa

Penguin Books Ltd, Registered Offices: 80 Strand, London, WC2R 0RL, England

First published by Penguin Group (NZ), 2006

1 3 5 7 9 10 8 6 4 2

Copyright © Val Monk, 2006
Copyright © illustrations remains with the individual copyright holders
The right of Val Monk to be identified as the author of this work in terms of section 96 of the Copyright Act 1994 is hereby asserted.

Designed by Athena Somerfeld
Prepress by Image Centre Ltd
Printed by Condor Production, Hong Kong

ISBN – 13: 978 0 14 302063 9
ISBN – 10: 0 14 302063 3

A catalogue record for this book is available from the National Library of New Zealand.

www.penguin.co.nz